Paleoalterites
and
Paleosols

IMPRINTS OF TERRESTRIAL PROCESSES
IN
SEDIMENTARY ROCKS

Paleoalterites and Paleosols

IMPRINTS OF TERRESTRIAL PROCESSES IN SEDIMENTARY ROCKS

Robert Meyer
Professor, University of Rouen
France

A.A. BALKEMA/ROTTERDAM/BROOKFIELD/1997

Published with the assistance of the Office of the Councellor for Cultural, Scientific and Technical Cooperation, Embassy of France, New Delhi.

Translation of : *PALÉOALTÉRITES ET PALÉOSOLS,* Editions du BRGM, Orleans, France, 1987. For the English edition update provided by the author in 1996.

© 1997, Copyright Reserved

Translators   : B.K. Srivastava and
                        Anuradha Banerji

Technical Editor   : Dr. S. Masood Ahmed

General Editor   : Margaret Majithia

ISBN 90 5410 724 3

Distributed in USA and Canada by : A.A. Balkema Publishers, Old Post Road, Brookfield, VT 05036, USA.

Foreword

Robert Meyer enumerates in a concise and practical manner the imprints of continental phenomena left on sedimentary and eruptive rocks. Weathered rocks, frequently encountered at the surface of continents, have been poorly studied. As these rocks are inconvenient for understanding the subsoil, they were ignored by geological mappers and mining geologists, who considered them 'ignoble'. However, over the years it became necessary to study weathered rocks because of their growing importance in geochemical prospecting, to ensure valid interpretation of anomalies, to explore material resources, and for management of urban and agricultural schemes. In view of these requirements, the study of surficial formations and their representations in geological maps have found favour.

Weathered rocks are formed by a complex interaction of many genetic processes, simple in themselves, requiring meticulous analysis, which Robert Meyer has succeeded in presenting very attractively, thanks to his experience, through the numerous examples illustrated in this book.

This book is the result of a successful collaboration with the Section of Surface Formations of the BRGM. Pierre Laville deserves special mention for his formulation of fundamental concepts and those pertaining to bauxite genesis. This manual is meant for geologists who have little formal training in pedology, rather than for pedologists. The author analyses how the products of surficial alteration are fossilised, how they respond to mechanical and chemical diagenesis, even to slow metamorphism, and their reaction on being exposed as an outcrop.

The examples should not be considered as an exhaustive study of pedological manifestations but rather as an initiation into utilising the tools of pedology. The proposed approach consists of the following distinct steps:

— To avoid, above all, hasty interpretation, even erroneous ones which discredit the authors.

— To learn to recognise apparently disorganised facies, quite unfamiliar to geologists, both in the field and in the laboratory.

— To envisage the possibility of identifying the different horizons and to reconstruct the profiles.

— To best utilise the interpretative elements of the model.

Technical and specialised terms have been kept to a minimum; unfortunately, total avoidance was not possible and hence a glossary has been provided at the end of the book as an aid to the readers.

C. Cavelier
Director of Publications, BRGM

Contents

FOREWORD	v
1. INTRODUCTION	1
1.1 History	1
1.2 Fundamental Concepts	2
2. EXAMPLES OF PALEOALTERITES AND PALEOSOLS	5
2.1 Biological Traces in Paleosols	5
2.1.1 *In-situ* Fossilised Plants	5
2.1.1.1 Plants associated with coal measures	5
2.1.1.2 Silicified plants	5
2.1.1.3 Root traces	6
2.1.2 Burrows in Continental Sediments	7
2.1.3 Horizons Rich in Organic Matter	8
2.1.4 Conclusions	9
2.2 Paleosols with Argillaceous Accumulation	9
2.2.1 Clay Minerals in Present-day Soils	9
2.2.2 Paleosols after Argillaceous Transformations	10
2.2.2.1 Examples of paleosol of Aquitaine molasse deposition	10
2.2.2.2 Criteria for identifying horizons of argillaceous accumulation	11
2.2.2.3 Elements of interpretation	13
2.2.3 Paleosols Rich in Neoformed Clays	13
2.2.3.1 Neoformed halloysite and kaolinite in the Wealden	13
2.2.3.2 Palygorskite horizon in the Miocene of Aquitaine	14
2.2.3.3 Diagnosis and elements of interpretation	14
2.2.4 Argillaceous Indicators in Paleosols	15
2.3 Carbonated Paleoalterites	15
2.3.1 Paleokarsts	16
2.3.1.1 Early and surficial paleokarsts	16
2.3.1.2 Primary paleokarsts	16
2.3.1.3 Secondary paleokarsts	17
2.3.1.4 Criteria for distinguishing between primary and secondary paleokarsts	18

	2.3.2	Tufas and Travertines		18
		2.3.2.1 Quaternary tufas and travertines		18
		2.3.2.2 Tufas and travertines in ancient deposits		19
		2.3.2.3 Conclusions		19
	2.3.3	Carbonated Crusts: Calcretes and Dolocretes		20
		2.3.3.1 Recent calcretes		20
		2.3.3.2 Fossilised calcretes in the Thanetian of Champagne		23
		2.3.3.3 Calcretes in ancient sequences		25
		2.3.3.4 Polyphased calcretes		25
		2.3.3.5 Fossilised dolocretes in the Permian of Saint-Die Basin (north-eastern France)		26
		2.3.3.6 Fossil dolocretes: Interpretations		30
		2.3.3.7 Accumulation of siderite		30
		2.3.3.8 Problems of detection and principles of interpretation		30
2.4	Sulfide or Sulfate Paleoalterites			32
	2.4.1	Present-day Sulfide and Sulfate Soils		32
	2.4.2	Paleosols with Gypsum and Anhydrite		33
		2.4.2.1 Fossil gypsum rosettes in the Ludian of Paris Basin		33
		2.4.2.2 Lenticular gypsum and anhydrite crystals		34
	2.4.3	Paleosols with Jarosite in the Lignites of Soissonnais		34
	2.4.4	Jarosite, Alunite and Barite in Paleoalterites		36
		2.4.4.1 Jarosite		36
		2.4.4.2 Alunite		37
		2.4.4.3 Barite		37
		2.4.4.4 Diagnosis of fossil mangroves		39
	2.4.5	Principles of Interpretation		39
2.5	Siliceous Paleoalterites			40
	2.5.1	Facts and Models		40
		2.5.1.1 Silica in continental environment		40
		2.5.1.2. Siliceous accumulations and 'recent' silcretes		40
	2.5.2	Alteration of Quartz at Paleosurfaces		41
		2.5.2.1 Splinter quartz		42
		2.5.2.2 Corroded quartz grains and desilicification		42
		2.5.2.3 Pebbles of weathered quartz		42
	2.5.3	Fossil Silcretes: Examples		43
		2.5.3.1 Silcrete in hydrolysing environment		43
		2.5.3.2 Silcrete in pre-evaporite environment		45
		2.5.3.3 Paleosequence with silcretes		48

			CONTENTS	ix
		2.5.3.4	Influence of burial diagenesis on silcretes	49
	2.5.4	Fundamental Types of Paleosilcretes		51
		2.5.4.1	Silicified conglomerates and sandstones	51
		2.5.4.2	Silicified clay	53
		2.5.4.3	Silicified calcretes and dolocretes	53
		2.5.4.4	Silicification in pre-evaporite environment	53
	2.5.5	Problems of Diagnosis of Fossil Silcretes		54
	2.5.6	Principles of Interpretation of Fossil Silcretes		56
2.6	Oxidised Paleoalterites			56
	2.6.1	Ferruginous Paleoalterites		56
		2.6.1.1	Iron in present-day alterites	56
		2.6.1.2	Rubefaction and formation of red beds	59
		2.6.1.3	Traces of hydromorphy fossilised in ancient sequences	59
		2.6.1.4	Paleosols with plinthites	60
		2.6.1.5	Ferruginous paleocuirasses	60
		2.6.1.6	Diagnosis and principles of interpretation	65
	2.6.2	Aluminous Paleoalterites		67
		2.6.2.1	Recent and present-day laterites	67
		2.6.2.2	Fossilised laterites in ancient sequences	69
		2.6.2.3	Bauxites	70
	2.6.3	Oxides and Hydroxides of Manganese		75
2.7	Paleosols on Volcanic Rocks			77
	2.7.1	Present-day Soils on Volcanic Material		77
	2.7.2	Examples of Paleosols on Volcanic Material		78
	2.7.3	Conclusions		79
2.8	Paleomantle of Alteration			79
	2.8.1	Examples of Paleoalterites on Hercynian Basement		79
		2.8.1.1	Triassic basement-cover contact in south-western Vosges	79
		2.8.1.2	Clay of basement and Triassic cover in Ardeche	81
		2.8.1.3	Albitisation of basement beneath a Liassic cover in Rouergue	81
	2.8.2	Precambrian Paleoalterites		84
		2.8.2.1	Limits of actualism	84
		2.8.2.2	Weathering profiles in the Precambrian of Saskatchewan (Canada)	84
	2.8.3	Diagnosis and Principles of Interpretation		86

3.	PALEOALTERITES AND PALEOSOLS IN THE TECTONO-SEDIMENTARY CONTEXT		87
	3.1 Factors Controlling Formation and Preservation of Paleoalterites		87
		3.1.1 Tectonic Context	87
		3.1.1.1 Paleoalterites and global tectonics	87
		3.1.1.2 Paleoalterites and sedimentary basins	88
		3.1.1.3 Paleoalterites and orogeny	88
		3.1.2 Climatic Context	89
		3.1.3 Topographic Context	90
		3.1.4 Geochemical Context	90
		3.1.5 Dynamics of Sedimentation	91
		3.1.5.1 Sedimento-pedogenetic cyclothem	91
		3.1.5.2 Preservation of paleoaltrites in alluvial sequences	91
		3.1.6 Formation, Development and Disappearance of Paleoalterites	93
	3.2 Specific Location of Paleoalterites in Geological Formations		94
		3.2.1 Paleoalterites Associated with Discordances	94
		3.2.2 Intercalated Paleoalterites in Sedimentary Sequences	96
		3.2.3 Polyphased Paleoalterites	96
		3.2.3.1 Bauxites on paleokarsts	97
		3.2.3.2 Outcrops of residual paleoalterites	97
		3.2.3.3 Iron hats	98
	3.3 Paleoalterites and Paleosols Through Geological Time		98
4.	LITHOGENIC ROLE OF CONTINENTAL ENVIRONMENT		100
	4.1 Transport of Material and Geomorphological Relief		100
	4.2 Geochemical Barriers		100
	4.3 Concentration of Minerals and Maturity of Sedimentary Rocks		101
	4.4. Chemico-Mineralogical Characteristics Acquired in Surficial Environments		101
	4.5 Physical Characteristics Acquired in Surficial Environments		102
5.	METHODS FOR THE STUDY OF PALEOALTERITES AND PALEOSOLS		105
	5.1 Study of the Outcrop		105
	5.2. Investigation of Paleoalterites		105
	5.3 Investigation of Paleosols		107
		5.3.1 Field Observations	108
		5.3.2 Microscopic Observation in Polarised Light	109
		5.3.2.1 Geological and pedological methods	109
		5.3.2.2 Specificity of pedological microstructures, convergence	110

		5.3.3	Mineralogy and Geochemistry	110
	5.4	Interpretations		110
		5.4.1	Interpretation of Paleoalterites	110
		5.4.2	Interpretation of Paleosols	111
			5.4.2.1 Reconstruction of profiles	111
			5.4.2.2 Inherent problems of paleosols	111
			5.4.2.3 Utilisation of pedological classification in geology	111
			5.4.2.4 Utilisation of diagnostic horizons	112
			5.4.2.5 Systematics of fossilised paleosols in ancient sequences	113
6.	KNOWLEDGE ACQUIRED FROM PALEOALTERITES AND PALEOSOLS			116
	6.1	Diagenetic Degradation of the Data Recorded in Paleoalterites		116
	6.2	Knowledge of Paleoclimate		117
		6.2.1	The Problems	117
		6.2.2	The Possibilities	117
	6.3	Knowledge of Duration		119
		6.3.1	Paleosols as a Unit of Time	119
		6.3.2	Paleoalterites as Stratigraphic Markers	122
	6.4	Knowledge of Paleogeography and Basin Dynamics		122
		6.4.1	Problem of Fossilisation	122
		6.4.2	Geochemical Paleoenvironments	123
		6.4.3	Knowledge of Tectonic Activity	123
	6.5	Appraisal		125
7.	GENERAL CONCLUSIONS			126
8.	GLOSSARY			127
	8.1	Some Important Soil Types		127
	8.2	Some Fundamental Pedological Micromorphology		128
		8.2.1	Transition from Macrostructure to Microstructure	128
		8.2.2	Fundamental Constituents	129
		8.2.3	Basic Fabric	129
		8.2.4	Plasma Separation	130
		8.2.5	Pedological Features	131
REFERENCES				133
INDEX				147

1. INTRODUCTION

Geologists have long correlated continental formations from various geological periods with well-classified environments, for example, 'feeder zones' of sediments or lakes in which limnic coals have been deposited. These concepts are fundamental and appear simplistic and stereotyped, giving the impression that the reconstitution of continental paleoenvironment came later than that of marine formations.

Multiple attempts to accurately reconstruct continental paleogeography on the basis of existing models have been undertaken in the last few decades. Such reconstructions facilitate the study of emergent surfaces, altered horizons and root traces. Geologists freely use the knowledge gained, particularly of *weathering processes* and *pedology*, as well as *geomorphology* and *botany*, both as investigating technics and for interpretation.

Pedologists had already made great advances in *paleopedology* by studying fossilised soils in all Quaternary formations. This led to the emergence of a new geological discipline, which may be termed the science of **paleoalteration** or *paleopedology*, which comprises the study of changes associated with emergent surfaces of all geological formations, including the most ancient. This discipline is the topic of the present work.

1.1 History

A complete historical analysis of the concepts of paleopedology emerges through the pages of this manual. Some landmarks illustrate the evolution of ideas in this domain.

Large discordant surfaces, often marked by erosional phases, have long been recognised but the presence of huge *in-situ* fossilised plants led geologists to the concept of fossil-soil, the most classical examples being the *Stigmaria* of coal measures or *in-situ* silicified trees.

An important landmark was the study of ancient continental formations, which were carefully analysed and described to reveal the specificity of their facies. For example, Vatan (1947) proposed microscopic descriptions of Tertiary continental deposits, which are entirely interpretable in terms of the **micromorphology** of soils. These interpretations were not favoured during Vatan's life, however, simply because micromorphology of soils was not in vogue, not even among pedologists. With similar ideas, but using different methodology, Millot (1949) distinguished continental and marine sediments based on clay mineralogy in the old series.

The geological and pedological schools of thought converged with the passage of time but Erhart (1956, 1967) must be credited with synthetising the ideas of the epoch in his theory of **biorhexistasy**. This model compares the periods of weathering (*biostasy*), during which continental rocks liberated only the most soluble cations, with the periods of **erosion** (*rhexistasy*), during which continental surfaces were eroded. Global interest in this theory made it a fundamental model; however, as it is not always applicable locally, misuse led to its discreditation.

Since the 1960s extensive studies have been carried out in two complementary directions:
— Following Millot (1964), extensive studies on the geochemical and mineralogical mechanisms in present-day and recent continental environments have been done, for

example by Lelong (1967), Paquet (1969), Tardy (1969), Ruellan (1970), Bocquier (1971), Souchier (1971), Nahon (1976) and Leprun (1979).

— Freytet (1964, 1970, 1973) emphasised the importance of detailed examination and interpretation of paleosols. Many others were of the same mind, for example Steinberg (1970), Buurman (1972), Haguenauer (1973), Watts (1977), Thiry (1981), Guendon and Parron (1983). Tardy (1993) proposed a very good framework of weathering mechanisms involved in tropical and/or paleotropical environments.

The same processes have been identified as active on widely varying formations. Their synthesis is the order of the day, either as joint endeavours, such as those edited by Yaalon (1971), Wright (1986) and Reinhardt and Sieglo (1988), or as individual efforts, such as Meyer (1981), Retallack (1981b), Freytet and Plaziat (1982).

1.2 Fundamental Concepts

Some fundamental expressions have been widely used in this manual. Their interpretations vary from author to author and hence it is necessary to define their meanings precisely as used here.

Alteration: *In-situ* transformation of a rock facies by the action of agents and under conditions that differ from those of its genesis.

Alterite: Product of mechanical and/or chemical *degradation* and/or *aggradation* of a rock. Some **transport** is implied, either of ions or of particulate matter. Alterology is the study of alterites and the mechanisms of alteration.

Meteoric Alterite: Alterites engendered by *atmospheric agents* (weathering), in particular water, which percolates through the rocks according to the laws of gravity, depending on their **porosity** and **permeability**. These alterites are generally loose, occasionally indurated; their mineral paragenesis is characteristic of *surficial thermodynamic conditions*, the chemistry of the 'parent rock' and/or source rocks, and *biological activity*. A meteoric alterite is easier to identify when it develops from endogenous **parent rocks** than when it derives from sedimentary rocks, because initially the characters may remain akin to those of the host rocks. In warm and humid climates alterites may be tens of metres thick, sometimes even more, depending on the depth of the water table. The alterite remains largely *in situ* or undergoes very little transport, which is not sufficient to give it a sedimentary organisation. Therefore, this term is not applicable to reworked material deposited in fluvial, lacustrine or marine environments.

Regolite (regolith): Meteoric alterite in which chemical weathering and aggradation are negligible compared to the processes of *physical disintegration*, such as thermal shocks or gelifraction.

Hydrothermal Alterite: Developed by *ascending hot waters* that traverse the rock according to its **porosity** and **permeability**. Its mineral paragenesis is related to the *thermodynamic conditions* and the *chemistry* of the *deeper zones* of the Earth's crust. The influence of the chemistry of the surrounding rocks is either absent or negligible. It is to be noted that if the fractures persist after the phase of hydrothermal alteration, **meteoric water** may circulate through them and induce a paragenesis of the meteoric type.

Profile of Alteration: The arrangement of facies of an alterite according to their gradient of evolution (Fig. 1).

Soil: Meteoric alterite with an asymmetrical, more or less vertical profile, and essentially related to a *pedogenesis* of significant *biological activity*. In the early stage there is

INTRODUCTION

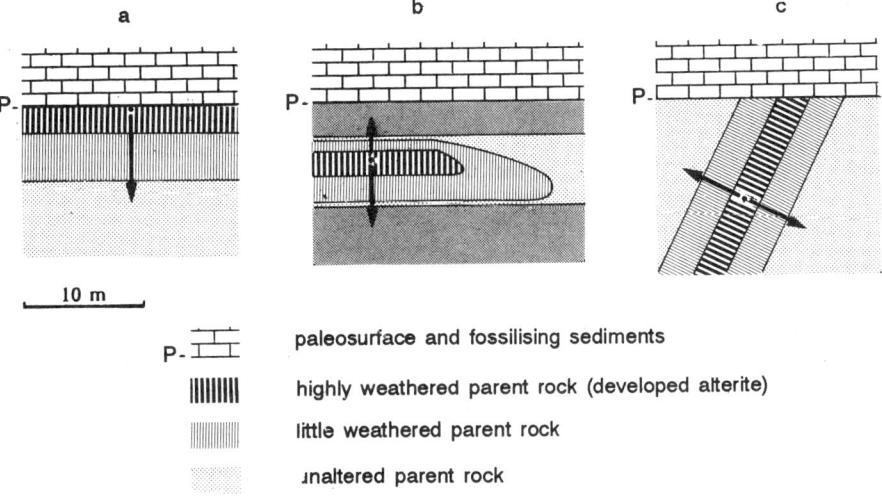

Fig. 1. *Classification of* **paleoprofiles of alteration** *based on their geometry. Arrows indicate the gradient from most altered facies to unaltered rock.*
a) Asymmetric paleoprofile under the paleosurface.
b) Dissymmetric paleoprofile developed particularly in permeable sedimentary layer.
c) Symmetric paleoprofile of hydrothermal origin.

a single thin horizon composed of primary minerals (Duchaufour, 1977) or **skeleton soil**. This progressively evolves into a differentiated soil in *several superposed horizons*. Pedology studies soils and the mechanisms of their genesis.

Parent Rock: Rock which after alteration produces an alterite or soil: This may undergo *complete transformation* of facies, rendering identification of the original rock difficult. It should be noted that the substrata of an alterite is not necessarily the parent rock (Fig. 2) and is not always the only source of elements found in the soil or the alterite.

Autochthony–Allochthony: Antonymous concepts indicating whether a *phenomenon* or *material* is *related* or *foreign* to an observed profile. When the material is related, the alteration phenomenon is necessarily related, but when the phenomenon is foreign, the affected material is also foreign. On the contrary, when a phenomenon is related to the profile, the parent rock may be foreign to that profile. These concepts, therefore, should be applied only with *reference to the material and/or the phenomenon*.

Paleoalterite and Paleosol[1]: The prefix 'paleo' attached to an alterite or soil indicates that the conditions responsible for its genesis are no longer those that prevailed earlier. Such facies are generally fossilised below younger rocks; if not, the alterites and soils are no longer in equilibrium with the ambient conditions and *tend to degrade*.

Diastem: Surface denoting a *short interruption* in sedimentation. This does not imply exundation but may be accompanied by erosion, capable of truncating a paleoalterite or paleosol.

---

[1] A small nuance in the language is the source of some ambiguity; in the strict French sense, paleosol should be logically translated in English as paleosoil, but the term is seldom used by Anglo-Saxon authors, who prefer paleosol, which is synonymous to paleosurface in French.

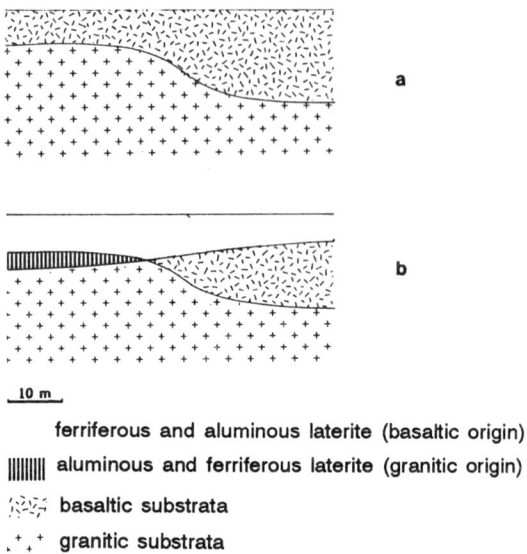

ferriferous and aluminous laterite (basaltic origin)
aluminous and ferriferous laterite (granitic origin)
basaltic substrata
granitic substrata

Fig. 2. *Vertical section of an alterite in equatorial humid climate.*
a) Section before weathering.
b) Section after weathering.

This example shows it is necessary to distinguish the parent rock, from which the alterite is derived from the substratum on which it rests. In general, alteration on **basalt** gives rise to a lateritic profile, thinner than on **granite**.

Paleosurface: A large surface of prolonged interruption in sedimentation. This interruption *implies an* **emergence**. The paleosurface is fossilised beneath younger rocks. Depending on whether erosion took place or not, it abuts unaltered rocks or paleoalterites.

Paleoprofile of Alteration: The paleoprofile can be conveniently classified according to the geometry of the phenomenon.

a) *Asymmetric paleoprofile* of meteoric origin, related to the *vertical percolation of surface water* (Fig. 1a);

b) *Dissymmetric paleoprofile* of meteoric origin related to *lateral percolation of surface water* (Fig. 1b):

c) *Symmetric paleoprofile* related to **hydrothermal percolations** through fractures (Fig. 1c).

In the case of meteoric weathered rocks, a schematic arrangement of diagrams is depicted in Fig. 1. It is to be noted that *the principle of superposition of beds is not applicable*: Fig. 1a shows the least altered and therefore the youngest facies at the base of the profile, while the oldest occurs at the top; in Fig. 1b, facies of the same degree of alteration and of the same age occur at two levels of the profile.

Superimposition of Alterites: During a single period of emergence and *a fortiori* a succession of emergences, frequently the same paleoprofile results from *several phases of alteration*, either of a similar or different nature. The resultant profiles are complex and can only be deciphered by mapping.

2. EXAMPLES OF PALEOALTERITES AND PALEOSOLS

The examples successively described here are diverse and complex; often the only common point is interpretation, namely, *rocks which carry the imprints left by meteoric agents*.

Field observations must necessarily be presented in an orderly manner. The cases presented here have been regrouped according to major processes, as follows,

— Fossilisation of *animal traces* or *vegetal remains*, if not present in present-day terrestrial deposits, prove biological activity, which was invariably present in the cases studied.

— The phenomena of **redistribution, migration** and **accumulation** of minerals such as argillaceous minerals, or chemical elements such as Ca, Mg, S, Si, Fe, Al, or Mn.

— Cases in which fossilised alterites exhibit no distinct specific characteristics, apparently due to the influence of a *particular substratum* (e.g. volcanic rocks), or even by **climatic** or **diagenetic conditions**, though poorly defined (e.g. Precambrian conditions).

The foregoing sequence has no particular significance and may even be disputed for lack of homogeneity. It has been adopted to initiate a novice into the subject.

2.1 Biological Traces in Paleosols

The remains of vertebrates and their footprints, are not only good stratigraphic markers, but also excellent indicators of **paleoenvironments**. The same applies to flora identified in coal deposits and to **pollens**. Other vestiges and traces of biological activity provide complementary information for the reconstruction of a continental paleoenvironment.

2.1.1 *In-situ* Fossilised Plants

2.1.1.1 Plants associated with coal measures
'**Fossil forests**' such as that of Champclauson found in the coal basin of Cevennes, with *in-situ* stems of *Syringodendron* (Vetter, Hery and Laversanne, 1975), are rare. This example shows that besides the roots and other subterranean parts, the aerial parts of large plants can also be fossilised, albeit not always preserved in the carbonaceous state but rather in the form of moulds filled with sandy material.

2.1.1.2 Silicified plants
The process of silicification usually affects small plants and is evident in acid sulfate soils (Buurman, van Breemen and Henstra, 1973). The silicified wood found in various geological formations is quite spectacular (Buurman, 1972; Beauchamp, 1981; Scurfield and Segnit, 1984). In fact, in most cases the wood is silicified after **transport** and deposition in alluvial formations. *In situ* **silicification** of plants is very rare. Recently, a controversy has developed over vertical silicified plant trunks in the 'Yellowstone Fossil Forests' (Eocene formation of Wyoming). Some studies indicate that the trunks were transported and

deposited randomly by mud flows as was observed during the recent eruption of Mount St. Helens (Fritz, 1980a, 1980b). Other studies provide very strong evidence in favour of *in-situ* silicification of plants which were rooted in poorly evolved soils (Retallack, 1981a; Yuretich, 1984). Some of these plants have 500 growth rings, thereby showing that the quiescence in sedimentation continued for many many years (Retallack, 1981b). These examples emphasise two important factors:

— Wood silicifies preferably in an environment rich with highly alterable rocks, such as **volcanic ash** (Meyer, 1984);

— Isolated vertical fossilised plant trunks may not be *in situ*.

2.1.1.3 Root traces

Among the present-day under-sea herbariums, only inferior plants or monocots are observed, which are feebly rooted and appear late in the history of plant life, probably proliferating during the Tertiary (Emberger, 1968). As for algae, it is doubtful whether their rhizoids can leave fossilised imprints. *Identification of root traces in pre-Cenozoic formations seems to imply an emergence, or at least a very thin water layer.*

Identification of traces of roots in older sequences is relatively easy, but care must be taken not to confuse them with burrows (Table 1). The term **rhizolite** is sometimes used (Klappa, 1980). If the medium is rich in organic matter and under reducing conditions, roots may fossilise to a carbonaceous state. If, on the contrary, the medium is less reducing, the roots disappear and only a deposit of iron oxides and hydroxides remains in the form of the original root. Such channels constitute the conduits around which various minerals are deposited. Thus, more or less vertical *carbonate casings,* sometimes empty along the axes (Freytet, 1971) and ferruginous casings are found in silicoclastic formations.

Table 1. *Criteria for distinguishing between burrows and roots (after Plaziat, 1971). These diagnostic features have only a statistical value*

Burrow	Root
Relatively constant diameter.	Highly irregular diameter.
Animals can perforate hard rocks.	Roots never perforate hard rocks (but a tender level can become hardened after perforation).
Internal structure often visible; striotubule with peripheral ornamentation.	Frequently fossilised into a hollow tubule.
Often filled early.	Often filled late.
Rare discontinuities due to intersections of burrows.	Truncated moulds; strings irregularly aligned.
Rare bifurcations, diameter unchanged, angles more often random.	Numerous ramifications, principal and secondary networks, angles generally defined and constant.

If properly preserved the network of roots can provide information about the **phreatic water table** that has disappeared. In sufficiently dry climates the network is essentially vertical in the **vadose** zone and has an extensive lateral spread in the phreatic zone and in the margins of marshes (Cohen, 1982; Mount and Cohen, 1984).

Root traces provide little information on the intensity of paleopedogenesis, as shown in the following example: At Cuis (Marne) a small Sparnacian (Lower Eocene) outcrop, two metres high and comprising seven to eight detrital sequences, is observed with distinct diastems. The top of each sequence is dark due to an increase in the percentage of carbonaceous remains (Fig. 3). Fine root traces are observed over the entire outcrop but they are too few to enable distinguishing one from the other. Some traces cut across more than one **sequence** and continue 30 to 40 cm vertically downwards. This root network developed without modifying the sedimentary structure, which indicates a rapidity and near absence of pedological differentiation and weathering. Thus *it would be foolhardy to attribute much paleopedological significance to* **root traces.**

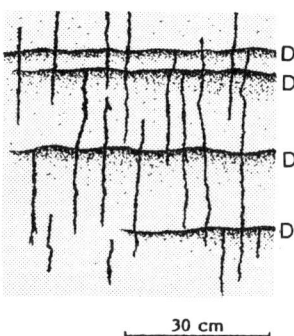

Fig. 3. **Root traces** *in the Sparnacian of Champagne. Fine oxidised tubules cutting across several detrital sequences without disturbing them, indicate the discrete character of pedogenesis (D: Diastem).*

In the Quaternary paleosols traces of roots are often inclined parallel to each other on the outcrop. This phenomenon may be associated with topographic undulations and interpreted as the sliding of surficial horizons, or **creep** (Yaadon, 1978). The phenomenon has been observed in older layers, e.g. barren Stephanian layers of Decazeville or argillaceous **'violet zones'** of Buntsandstein in north-east France (Durand and Meyer, 1982). Root traces measuring more than a metre in length have a common inclination of approximately 30° with the vertical. Traces of fossilised roots in the pelitic layers may develop a characteristic *accordion* form, since the compaction is considerable. Diagenetic deformation of fossil root traces is common.

2.1.2 Burrows in Continental Sediments

Burrowing animals in the exundated sediments produce a more rapid mixing of the soil than do plants. A soil horizon rich in **earthworms** can be completely processed through the digestive tubes of the worms in five years (G. Aubert, pers. comm.). The animals found in present-day soils are numerous and varied; the scenario must have been much the same in earlier times. **Termites,** in particular, must have played a significant role (Alidou, Lang and Lucas, 1977).

In environments where land and ocean converge, identification of burrows is of interest only if they can be differentiated as marine or continental, which is a delicate problem. Plaziat (1971) emphasises the importance of differentiating between burrows with an ornamental surface of the type ophiomorph, abundant in littoral marine environments, and continental burrows with an imbricate structure in striotubules (Freytet and Plaziat, 1982, p. 204). Although **striotubules** are sometimes found in a marine environment, they are in fact more abundant and relatively characteristic of a continental environment (Allen Curran, 1985). A few examples of continental burrows are depicted in Fig. 4.

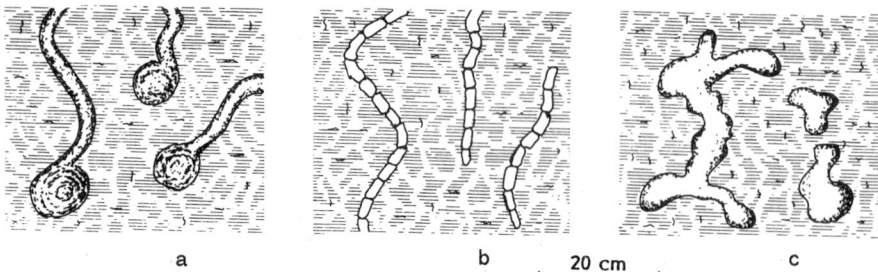

Fig. 4. **Burrows** *fossilised in continental formations, observed in vertical sections.*

a) Burrows of earthworms which terminate in a chamber; fine black deposits of manganese oxide emphasise a lamination especially apparent in the chambers. Continental Stampian molasse, Aquitaine.

b) Burrows probably formed by earthworms, filled with micrite, with no internal structure or central void. Continental Stampian molasse, Aquitaine.

c) Burrows of small mammals in loessic sierozem hidden under other soils; they are at least 25,000 yrs old. Netivot, Israel.

2.1.3 Horizons Rich in Organic Matter

Surficial horizons of soils are generally eroded before fossilisation occurs. However, **histic horizons** may be fossilised under periglacial deposits (Schneebeli, 1976), or under **aeolian silts**. If these layers are not destroyed by **oxidation,** they undergo considerable compaction in the older sediments which reduces them to simple laminae. Identification is then difficult and organic geochemistry becomes the key method, as long as the sampling is closely spaced. *Oxidation tends to degrade* the organic molecules, at least partially, to give relatively short alkane chains (Didyk et al., 1978). The composition of sterols in the sediments also provides information on the biological sources of their genesis (Huang and Meinschein, 1976; Tissot and Welte, 1978).

Organic matter may also accumulate in the deeper soil horizons (horizon Bh), as shown in the following examples: Williams (1968) discussed the possibility of a 1000 Ma old paleoalterite from Scotland, which might have undergone a podzolic type of evolution. Grey and Grandstaff (1980) have interpreted some paleosols in Canada of 2300 Ma as *podzols* quite rich in organic matter (0.25%). Ellenberger, Feys and Trichet (1967) observed layers rich in organic matter at the top of the Fontainebleau Sands (Stampian), which could have resulted from reworking of the Bh horizon of podzols: organic matter is in fact, polymerised in humic acid and has a high carbon-nitrogen ratio (C/N = 20). Pomerol (1964) described a succession of horizons, some of them rich in organic matter, overlying the Auversian sands of the Paris Basin, which he interpreted as humic paleopodzols. In the quarry of Fere-en-Tardenois (Aisne), the sands comprise dark bands rich in organic matter, present as small granules a few tens of microns in size, which do not adhere to the unweathered quartz grains. This is characteristic of **spodic horizons** of podzols (Duchaufour, 1977, p. 331) but it has been observed that organic accumulation follows the foliations of stratification and the material seems to be reworked rather than *in situ*.

It is no less true that *podzolisation has been able to leave traces in older sequences;* the soluble organo-mineral complexes tend to protect organic matter from biodegradation.

2.1.4 Conclusions

— Plants, large and small, may undergo varied **epigenesis** which fossilises them. They cannot be interpreted in paleopedological terms until proved to have grown *in situ.*

— Root traces are a good index of **emergence,** but since they develop rapidly are not very significant in pedological differentiation.

— The fauna that live beneath the soil leave **burrows,** which can be distinguished from root traces as well as marine burrows.

— Humic matter, when fossilised, retains at least at some stage of its **diagenesis** the memory of its continental origin.

CONTINENTAL BIOLOGICAL TRACES

Characteristic Morphology:
Burrows: elongated form, relatively uniform sections.
Roots : elongated form, dichotomous, irregular sections.

Microfacies:
Striotubules for the burrows, cellular structures often visible in fossil wood.

Mineralogy:
Common epigenesis, particularly of vegetal remains by silica, phosphates, sulfides, etc.

Substrata:
Roots and continental burrows traverse only friable substrata.

Frequency:
Common at least since the base of the Paleozoic era.

Paleoenvironment:
Root traces indicate a very thin water cover and even very often an exundation, but probably of short duration (one or several years).

2.2 Paleosols with Argillaceous Accumulation

Argillaceous minerals accurately record transformations of surficial environments. Two approaches based on the use of landscapes and profiles may be envisaged for investigating ancient sediments.

The following are some examples:

2.2.1 Clay Minerals in Present-day Soils

Migrations of clay minerals, which take place essentially in a complex argillo-humic state (Souchier, 1981), can be effective only in the absence of carbonates or after their **leaching** (Duchaufour, 1977). This is manifested by relative enrichment of the **B horizon** with clay compared to the **A horizon** or **C horizon** (this enrichment cannot take place unless the source material is homogeneous). **Argillans** and **ferri-argillans** are characteristic of these enriched horizons; they produce **illuviation** at the microscopic scale (Pedro, 1993).

In soils associated with warm climates and considerable **drainage,** alteration of clay minerals is expressed by a destabilisation, the corollary of which is **neoformation** of phyllosilicates of the type 1/1 (kaolinite), or even hydroxides in a highly hydrolysing medium. Topographic depressions and poor drainage cause the formation of **vertisols** and neoformations of **smectite**, whereby accumulation occurs downstream upwards (ascending invasion; Paquet, 1969). Thus in these types of topography, in tropical regions, where the climate is contrasting and sufficiently humid (500–800 mm/annum), a veritable *smectite cover* may form both on endogenous rocks as well as on pelites or limestones (Millot, 1982).

In temperate climates alteration is more restrained. The primary phyllosilicates undergo a *microdivision* giving rise to illite, flakes of which in turn *degrade to smectite* or *vermiculite* (dioctahedral). Neoformations exist but on a restricted scale (Duchaufour, 1977).

Table 2 gives some chemical distribution patterns of clay minerals in present-clay soils. It is necessary to study them for comparisons while investigating paleosols. In fact, a moderate diagenesis may influence the mineralogy of clay without appreciably transforming the chemistry of the facies.

Table 2. *Chemical composition of some comon clay minerals in soils of continental environment. The average ratios (molar) are calculated on the basis of data provided by Weaver and Pollard (1973). It is an indicative value for the purpose of comparing different minerals*

	SiO_2/Al_2O_3	Al_2O_3, K_2O, MgO
Kaolinites	2.0	$K_2O + MgO < 10^{-2}$ mole/100 g
Sericites	2.6	$Al_2O/K_2O \approx 3.4$
Illites	3.2	$Al_2O_3/K_2O \approx 3.5$
Vermiculites of soil	3.7	$MgO/Al_2O_3 \approx 0.9$
Smectites (Montmo.-Beidel.)	4.6	$MgO/Al_2O_3 \approx 0.4$
Palygorskites	7.3	$MgO/Al_2O_3 \approx 1.7$
Sepiolites	64	$MgO/Al_2O_3 \approx 36$

2.2.2 Paleosols after Argillaceous Transformations

2.2.2.1 Examples of paleosol of Aquitaine molasse deposition

The poor preservation of surficial horizons of paleosols and the heterogeneity of **parent rocks** make it difficult to pinpoint the horizons of clay accumulations. However, examples exist from the Stampian of Central Aquitaine. Continental molasse is a slightly carbonated clay-sandy sediment. This molasse contains several paleosols with comparable profiles. The evolution of one of these paleosols is systematically depicted in Fig. 5 (Meyer and Guillet, 1980).

— The sediments deposited in the piedmont zone, occur above the **phreatic water table.** Drainage of **meteoric waters** through them causes the growth of vegetation, which leaves radicular tubules.

— Pedogenesis commences; it is affected by the migration of carbonates which accumulate at depth; **chlorite** and **biotite** simultaneously start to alter and split into smaller crystals in the upper part of the profile.

— When the surficial horizon is completely decarbonated, alternate drying and moistening imparts a polyhedral structure to it. The polyhedral faces and the tubules become

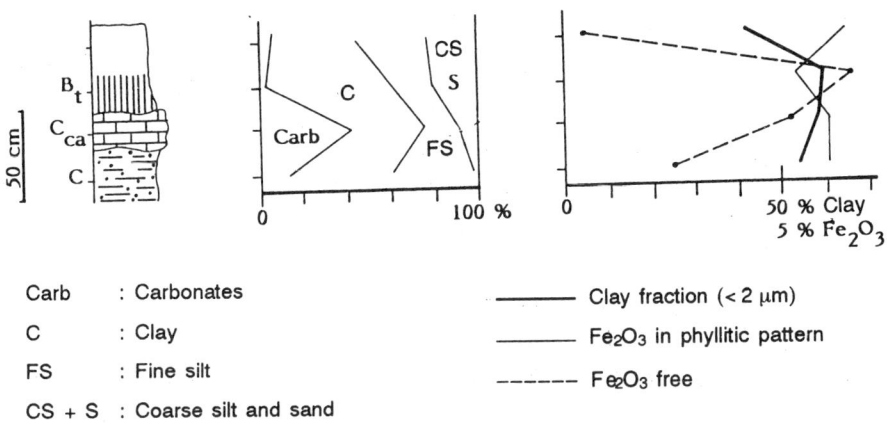

Carb	: Carbonates
C	: Clay
FS	: Fine silt
CS + S	: Coarse silt and sand

— Clay fraction (< 2 μm)
— Fe_2O_3 in phyllitic pattern
------ Fe_2O_3 free

Fig. 5. *Paleosol in the Stampian molasse of Aquitaine. Lithological section and composition of horizons. The diagram on the extreme right summarises the composition at the thin phase: percentage of clay minerals of iron in silicate structures and of free iron.*

the planes of *joint migration of clay and iron hydroxides*. A B_t horizon appears, which is red in colour. Under the microscope many **ferri-argillans** and **papules** which resulted from reworking are clearly visible (Fig. 6a, b). It should be noted that the clay resulting from mechanical leaching is but a small percentage of the global clay occurrence (Fig. 5).

— The subsequent transformations, which are no longer pedological strictly speaking, could complicate the profile, in particular the surficial part which turns greenish under the effect of an acidic and reducing medium, probably associated with degradation of organic matter during the arrival of fresh sediments which seal the profile.

2.2.2.2 Criteria for identifying horizons of argillaceous accumulation

A quantitative estimation of the clay percentage in each horizon of a profile, repeated on several profiles, is certainly the most reliable method (Meyer and Guillet, 1980). The method is very tedious and only applicable to paleosols not hardened by diagenesis.

— The **argillans** and **ferri-argillans**, because of their special characteristics under the microscope, are good indexes of illuviation. They are well fossilised, even though hydromorphic environments may degrade them (Brinkman et al., 1973). They have been widely interpreted by geologists (Teruggi and Andreis, 1971; Ducloux, 1973; Meyer, 1976; Daugas, 1981). It is prudent to use them in conjunction with other indexes, since they are likely to be associated with non-pedological processes (Brewer and Haldane, 1957) and more often with recent pollutions, because sampling is done at the outcrop or just beneath the surface.

— **Papules,** which are the result of reworked *in-situ* argillans and ferri-argillans of a profile, are more reliable indicators than argillans; their deposition precludes them from recent pollution.

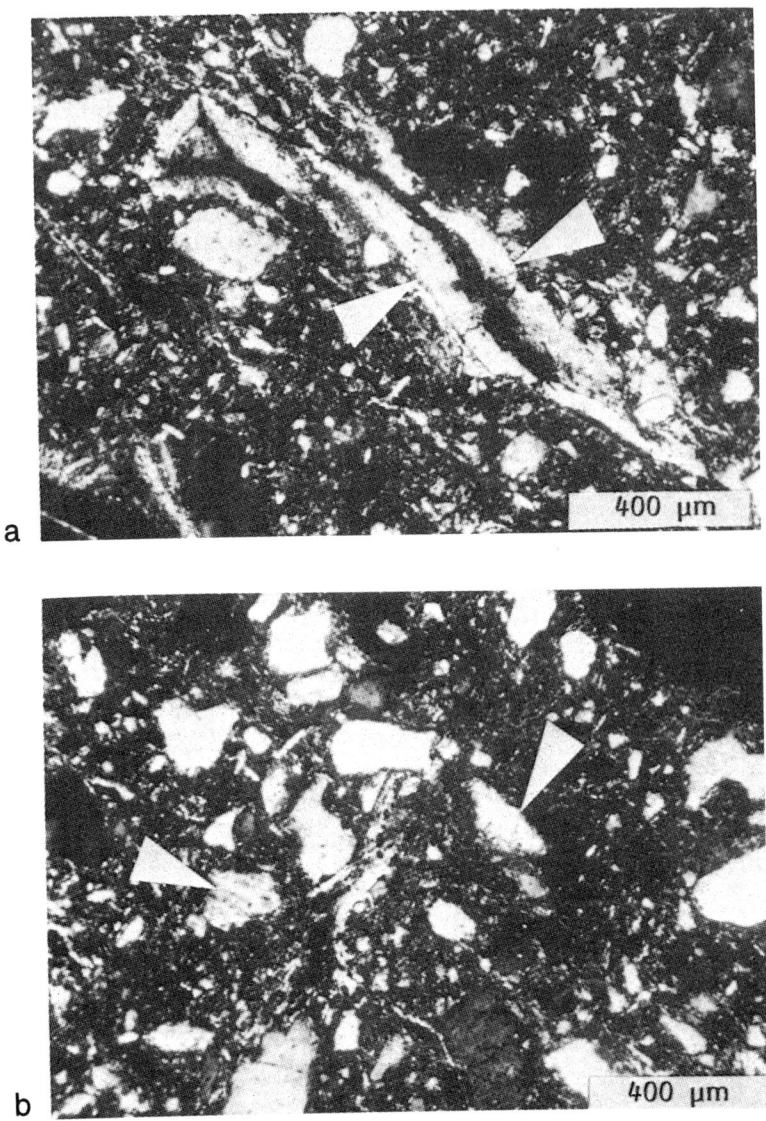

Fig. 6. *Characteristic microstructure of B_t horizons (microscopic view in polarised light).*

a) Argillaceous particles are intimately mixed with iron-hydroxides constituting **ferri-argillans** *which line the faces of* **pedotubules**; *all clay particles have a common optical orientation and are orange in natural light, with parallel to polarizer extinction in polarised light.*

b) The **plasma** *of a B_t horizon, dark in colour encloses quartz grains and clay aggregates (flakes) derived from reworking of ferri-argillans (papules).*

— Global variation of chemical composition of horizons is also a criterion, which should be considered a function of the clay percentage of a horizon and the composition of clay minerals (see Table 2).

2.2.2.3 Elements of interpretation

An **horizon** rich in clay formed by leaching in a biologically active environment represents a *fairly long period* of evolution (hundreds or thousands of years), since it must have been preceded by leaching of carbonates, if present.

In an alluvial sedimentary formation the well-preserved clay argillans and ferri-argillans imply a feeble **hydromorphy** in soil (Brinkman et al., 1973), a relief in the alluvial plain and a certain pattern of watercourse migrations, which allow sediments to stay above the water table for long periods (Rocha and Gomez, 1992).

2.2.3 Paleosols Rich in Neoformed Clays

Neoformation is always complementary to the alteration of other minerals; the following examples are no exceptions to the rule.

2.2.3.1 Neoformed halloysite and kaolinite in the Wealden

The Valanginian of the Wealden facies in north-eastern France (Meyer, 1976) consists of sandy-clay and reveals the pedological paleoprofiles shown in Fig. 7. These comprise root traces, nodules of grey clay and ferruginous pellicules occupying the voids. The **plasma** presents a **skelsepic** sometimes **vosepic** microstructure; minerals of the kaolinite family increase towards the top of the profile at the cost of illite which decreases in percentage. In this formation, the long halloysite crystals are characteristic of paleosols; their form excludes the possibility of a detrital origin (Fig. 8a).

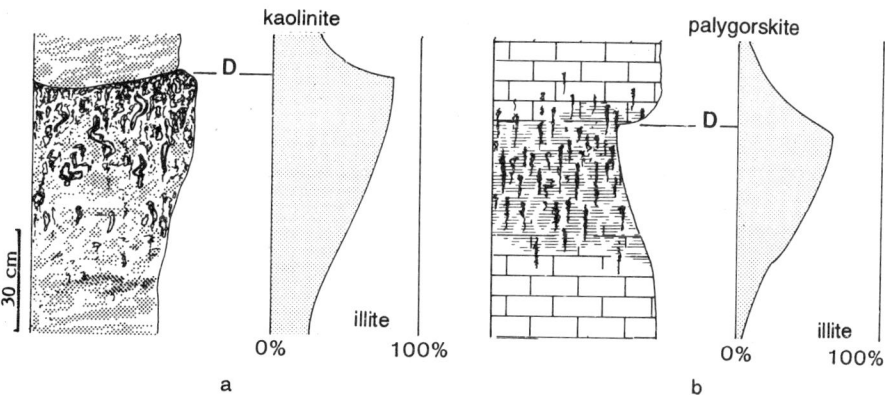

Fig. 7. *Evolution of a clay assemblage in two paleosols.*

  a) Wealden in north-eastern France: The maximum percentage of kaolinite appears at the base of diastem D, which represents a slow erosion before the recurrence of sedimentation.

  b) Miocene in Aquitaine of south-western France. Paleosol rich in palygorskite within lacustrine limestone; there is no distinct erosion at the top of the paleosol.

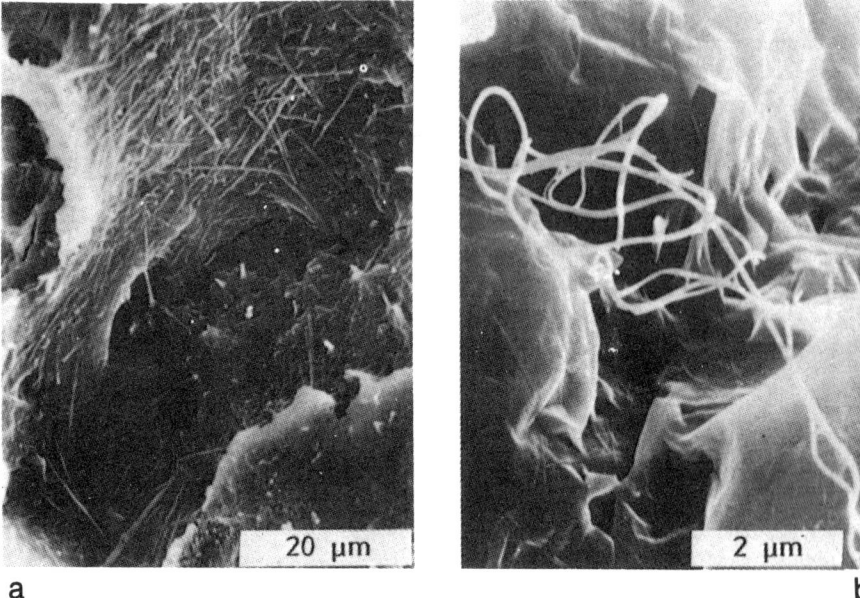

Fig. 8. *Fibrous argillaceous minerals whose properties under the SEM prove authigenesis.*
  a) **Halloysite** *filling a pore in the Wealden paleosol.*
  b) **Palygorskite** *fibres in paleosol of the Miocene in Aquitaine.*

2.2.3.2 Palygorskite horizon in the Miocene of Aquitaine

In the alluvial plain where this molasse is deposited, a micritic sedimentation invades the zones occupied by shallow lakes (Crouzel and Meyer, 1977a). Desiccation seems to be the origin of green clayey paleosols within limestones (Fig. 7b). The clayey layer is highly disturbed by *root traces,* consisting of up to 60% palygorskite, whereas this mineral is absent from **lacustrine limestones** which consist of only illite and a little smectite. The palygorskite characteristic of this horizon is apparently **authigenic** (neoformed) as confirmed by observations under the scanning electron microscope (SEM) (Fig. 8b). Magnesium, derived from primary phyllites **(chlorites)**, necessary for authigenesis, probably goes into solution in the highly basic water. This palygorskite facies is again found as reworked pellets in overlying limestones.

2.2.3.3 Diagnosis and elements of interpretation

A geochemical approach to the mineralogical balance of a profile accounts for argillaceous authigenesis. Clay particles when viewed under an electron microscope provide additional information.

Neoformation of kaolinite resulting from the alteration of illite is *a slow process,* probably requiring thousands of years. The *neoformation of palygorskite* in an evaporite environment might be *faster,* which is confirmed by the character of the diffused diastem at the top of the paleosol (Fig. 7b).

2.2.4 Argillaceous Indicators in Paleosols

Clays appear to be *very sensitive to pedological evolutions* and are consequently good indicators. Besides the simple mechanical reworking or the authigenesis described earlier, it is necessary to note the vast range of clay minerals present in paleosols: almost all may play a significant role in these environments, sometimes very markedly specific. They include, for example, ferriferous kaolinite (Cantinolle et al., 1984), ferriferous illite close to glauconite (Parry and Reeves, 1968; Durand, 1975), aluminous smectite (Thiry, 1981, p. 45) or the interstratified clays such as 7–14 sm, recognised particularly at the base of Tertiary sediments of the Paris Basin (Laurain and Meyer, 1979). Robinson and Wright (1987) proposed interpretations of clay minerals in Carboniferous paleosol sequences.

HORIZONS ENRICHED IN CLAY

Colour:
Brown to red, sometimes grey to green.

Microfacies:
Argillans, ferri-argillans, papules, stress cutans

Mineralogy:
Kaolinite, illite, smectite, vermiculite, palygorskite and others; these minerals naturally evolve during burial diagenesis.

Substratum:
Necessarily silico-aluminous, since it is the parent rock; varied if the clays are authigenic due to solutions foreign to the profile.

Frequency:
Common in all epochs.

Dominant climatic conditions
Kaolinite   : Warm and relatively humid;
Illite   : Temperate;
Palygorskite : Dry pre-evaporite environment

Convergence:
Accumulation of clay by mechanical sedimentation, where the layers are regular and better stratified.

2.3 Carbonated Paleoalterites

The mobilisation and fixation of **carbonates** in continental environments are processes opposed to each other at some point but complementary in the context of the overall landscape; they are occasionally readily detected in the same paleogeographic reconstruction. Mobilisation of carbonates takes place at all scales; *paleokarsts* are the final form of development of this mobilisation and examples are described under separate headings. Carbonate accumulations occur in their various forms, from *tufa* and *travertine* to different types of *calcretes*. Finally, the formation of dolomite and siderite will be considered as a function of surficial environments and diagenetic conditions.

2.3.1 Paleokarsts

Present-day karstic topography is often the result of long periods of evolution, which are not operative today, but appear as simple relics. Geological literature includes such karstic paleosurfaces associated with past and fossilised landscapes found in geological sequences. Water dynamics in present-day karstic systems is hard to understand and is not totally known (Vervier, 1990).

Karstic morphology can develop on diverse rock types, provided they are relatively soluble in the climatic conditions in which they outcrop. *Carbonated rocks* are the most frequently mentioned, but karst may also develop from *gypsum* (Toulemont, 1984), *anhydrite* and *salt, basalt* or *peridotite,* such as the *dunites* of New Calendonia (Trescases, 1975), or even *sandstones* (Urbani, 1978). It is relevant to mention here that karsts can also develop under cover. Thus limestone may become karstified when covered by a sandy formation or sandstone, which is eventually disturbed by subjacent dissolutions. This poses a difficulty in establishing the relationship between the cause and the effect. Sinkhole fields now covering the Carboniferous Grit Millstone of Wales are a good example. Some authors consider this as karstification enveloped by sandy formations (Battiau-Queney, 1984), while others describe it as a simple consequence of karstification in subjacent limestone (Wright, 1982a).

Identification of paleokarsts is of some importance. Besides the paleogeography, paleoclimate and geodynamic reconstructions (Guendon, 1984), they also constitute the substratum of many **bauxite** deposits and are often the preferred horizon for *lead-zinc* concentration (Bernard, 1972; Lagny, 1974; Melas, 1982), for *phosphorite* (Billaud, 1982), and even for petroleum.

Combes (1978a, b) and Melas (1982) distinguished three principal types of calcareous paleokarsts based on their genesis. These three types may be linked in time; in particular surficial karsts may be replaced and hidden by primary karsts or even secondary karstification.

2.3.1.1 Early and surficial paleokarsts
Associated with the first phase of emergence, this type of paleokarst may develop from a few centimetres to several metres. It is characterised by a network of dense dissolution cavities (Fig. 9a), breccia and small slumps. The dissolution cavities undergo geopetal filling of the vadose-silt type, subsequently followed by occlusion by sparitic calcite (Fig. 9b).

Such early paleokarsts develop in lithified limestone but the emergence of carbonated mud may give a similar structure, especially at the stage of dissolution cavities; this occurred, for example, in the shallow lakes of the Miocene, Aquitaine (Crouzel and Meyer, 1977a), and in the Eocene marshes of Languedoc (Plaziat and Freytet, 1978).

2.3.1.2 Primary paleokarsts
Immediately after a general **emergence,** the vertical amplitude became a function of the height of the exundated rock (Esteban and Klappa, 1983). It is characterised by an irregular surface at the metric or decametric scale, with pinnacles, sinkholes and large dissolution cavities and breccias (Fig. 9c).

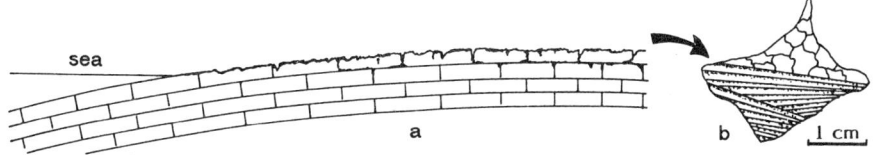

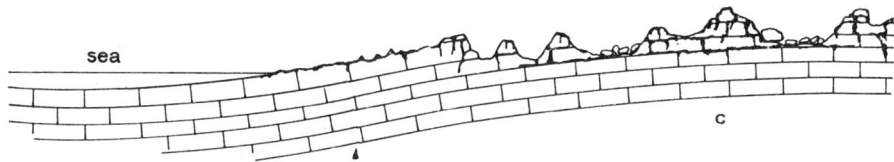

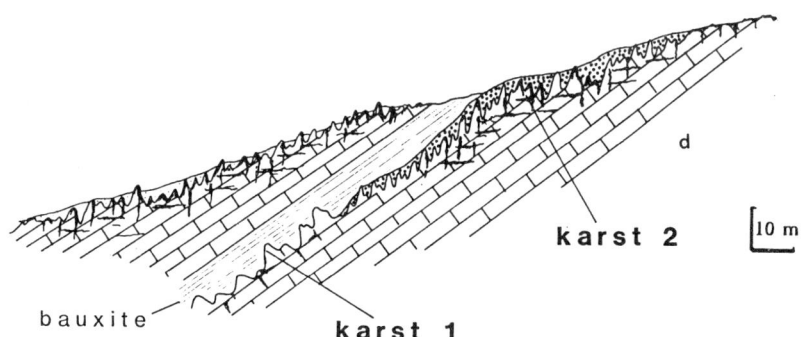

Fig. 9. *Diverse stages of evolution of a* **karst** *as preserved by fossilisation.*

a) Early karst developing on a feebly emerged zone.

b) Dissolution cavities may be partially filled with vadose-silt deposited as fine laminae, the upper part of the cavity filled with sparite.

c) Well-developed primary karst.

d) Karst associated with **bauxite** *deposits (after Combes, 1978a). The bauxite occurs on top of the primary karst; a post-tectonic secondary karst develops under the cover of bauxite breccia or terra rossa.*

2.3.1.3 Secondary paleokarsts

Much later, secondary paleokarst which corresponds to *dissolution under cover* develops, with certain implications:

— Its dating is often difficult.

— It does not imply emergence at its immediate contact, which is an important factor in the definition of regression—transgression cycles (Guendon, 1981).

— An ancient secondary paleokarst may be confused with Quaternary alterations; in fact, meteoric waters circulate preferably through the joints of stratification, which gives rise to dissolution cavities of the karstic type.

2.3.1.4 Criteria for distinguishing between primary and secondary paleokarsts
The tectonic criterion is excellent, when it exists, for differentiating between primary and secondary paleokarsts. Dissolution conduits in active karsts are essentially vertical and horizontal; if a secondary evolution is separated from the primary by a tectonic phase, a new orientation of pinnacles is superimposed on the primary (Fig. 9d).

Wright (1982a) described a few features specific to primary paleokarsts. The absence of these characteristics turns the interpretation towards a secondary paleokarst:

— A primary paleokarst *may be covered by a paleosol* or an alluvial formation and subsequently be sealed itself by later deposits (Chafetz, 1982).

— The paleokarst may be *partially eroded* at the contact with the overlying sediments.

— No lithological continuity, not even local, should exist between the paleokarst and the overlying sediments.

— *No breccia* of the constituents of *overlying* lithified formations should be present in the karst.

2.3.2 Calc Tufas and Travertines

Tufas and travertines are calcareous concretions. The precipitation responsible for their formation could be physicochemical but more often they are the result of the biological activity of *bacteria, cyanophores* and *bryophytes*. They are known to contain *casts of the upper parts of plants* (leaves, stems etc.).

Calcareous tufas are *massive, porous and spongy;* travertines are *laminated and compact.* Geologically, their distinction is of limited value because the interpretation of these terms varies from author to author, especially since tufas and travertines often occur intermingled in the same formation.

2.3.2.1 Quaternary tufas and travertines
Streams flowing on calcareous terrains, however small they may be, deposit travertines that coat their beds. This phenomenon occurs in all climates but extends over vast regions in *Mediterranean countries*. Travertines also occur in **subarid climates** but are confined to the rare zones where surface water flows; elsewhere, they are replaced by calcareous crusts. **Hydrothermal water** may lead to large deposits of travertines (Lang, 1975, p. 530; Chafetz and Folk, 1984). The thermal anomaly may be interpreted as tectonic conditions taking precedence over climate and elevation. These latter forms are recognised occasionally as concretions with ascending growth, developing above the level of emergence.

Facies vary with diverse types of travertines (Fig. 10): *cascade* facies are generally spongy, abundant in large internal cavities and ultimately include *stalactites*, while forms on the bed of streams or lakes are more akin to laminated **stromatolites** and only slightly porous (Anadon and Zamarreno, 1981). Adolphe (1981), Casanova (1981), Guendon and Vaudour (1981), Chafetz and Folk (1984) discuss in detail the role played by organic activity in the development of these structures, in particular the role of bacteria, their seasonal growth and the relationship between constructing organisms and petrology.

EXAMPLES OF PALEOALTERITES AND PALEOSOLS

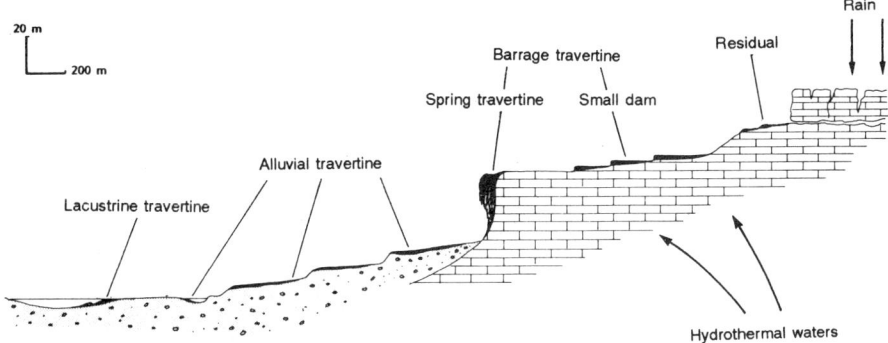

Fig. 10. *Synthetic section showing diverse types of* **travertines** *recognised in present-day landscapes. Naturally the lowest part of the landform has the maximum probability of fossilisation. Alluvial travertines may seal ancient terraces or encroach on active watercourses. Without being necessary, hydrothermal waters may contribute to extensive development of travertines in a region, due to their load of calcium bicarbonate and their temperature.*

2.3.2.2 Tufas and travertines in ancient deposits

Oligo-Miocene lacustrine deposits of Limagne (central France) are rich in **stromatolitic** and oncolitic facies, which have been studied in detail (Donsimoni and Giot, 1977). Near the calm shores of the lakes huge masses of travertines with bryophytes or phryganeid larvae have been consolidated by stromatolitic deposits; this assemblage may have formed a veritable **lacustrine reef** that fossilised (Freytet and Plaziat, 1982, p. 40).

Another classical example is the *travertine of Sezanne,* dated from plant imprints as Upper Thanetian (Lower Eocene, Paris Basin). These imprints have helped in reconstructing the flora that grew in the environment of the deposit, but the possibility of a faulty **paleoclimatic** interpretation cannot be ruled out. The floraflourished in a relatively humid climate, which does not necessarily imply heavy rains but indicates, at most, the presence of local watercourses necessary for the formation of travertines. Recent studies (Laurain and Meyer, 1979, 1984, 1986) highlight certain points. A similar formation of travertine, contemporary to that of Sezanne, has been found to occur locally at several places, from the slopes of Champagne up to the Reims Mountains. *Laterally it grades into* **calcrete** *crusts* (see Fig. 13) or to poorly cemented calcarenites, consisting of debris from the travertines. These diverse facies can be fitted into a coherent paleogeographic model: in a relatively **arid** climate calcretes develop on higher reliefs; at a lower zone where water circulates almost permanently, travertines are developed, the *partial erosion of which results in calcarenites;* their weak cementation probably expresses lengthy periods of desiccation of the alluvial beds in which they were deposited.

2.3.2.3 Conclusions

Tufas and travertines, which are the products of physicochemical conditions or biological activity, could fossilise in ancient deposits. Besides confirming a continental origin, they provide information on the paleoenvironment:
— *Hydrothermal activity* which facilitates their development;
— *Specific* **relief**, necessary for the appearance of cascade and retentive forms;

— **Mediterranean** *climate*, which favours regional extension;
— *Paleoflora* through paleobotany.

TUFAS AND TRAVERTINES

Characteristic morphology:
Foreign to substratum; discontinuous formation in vertical extension (mainly cascade tufas) or horizontal extension (mainly river or lake travertines).

Characteristic facies:
Vegetal casts, stromatolites, oncolites,

Microfacies:
Essentially micritic.

Climate:
Mainly Mediterranean and similar climates.

Diagenesis:
Initial high porosity diminishing rapidly with cementation during burial.

2.3.3 Carbonated Crusts: Calcretes and Dolocretes

The term **calcrete** has been used to designate many different entities and it is difficult to attribute a precise meaning to it. The following definition has been accepted here: a calcareous concretion which develops in soil or in alterites; the accumulation appears as a continuous horizon or an endurated band. This definition excludes **lacustrine** limestones, which commonly have specific characteristics. However, this definition is not universally accepted. For example, the uraniferous calcretes of Australia are essentially lacustrine (Briot and Fuchs, 1978; Carlisle, 1978).

The term **caliche**, used earlier, is synonymous with calcrete. However, calcrete is now being internationally accepted because similar etymology (English root: concrete) is used to denote other accumulations of continental milieu: **dolocrete** (dolomite), **silcrete** (silica), **ferricrete** (iron oxides), **alcrete** (aluminium hydroxide), **gypcrete** (gypsum) etc.

2.3.3.1 Recent calcretes
An enormous bibliography exists on Quaternary calcretes. Goudie (1973, 1985) and Reeves (1976) have proposed a global review. Some fundamental points valid for the study of recent calcretes are useful in the investigation of ancient deposits.

— *Climate:* Calcareous concretions appear in various climates, even in a cold climate (Sweet, 1974), but the phenomenon is extended only in warm and relatively dry climates, such as the Mediterranean, subdesert and desert. The composition of stable **isotopes** of C and O should remain in equilibrium with the climate prevailing during their formation (Gerling, 1984).

— *Substratum:* It may vary from calcareous or marly rocks to rocks devoid of limestone but containing the minerals susceptible to liberating calcium during weathering (plagioclases). Calcretes are even found on rocks devoid of calcium, which means an *external transport* (**aeolian**), and a slow rate of formation.

— *Thickness*: Varies from a few centimetres to several metres.
— *Colour*: White, creamy, sometimes coloured by oxides of iron (salmon crust).
— *Morphology*: Some calcretes are powdery, though they are mainly massive with no particular structure. The well-developed forms of calcretes (Ruellan, 1970; Arakel, 1982) present a succession of distinct characteristic horizons (Fig. 11); **banded pellicules** rest on top of **flagstones** or a very hard **cuirasse,** which in turn overlies a more porous **crust. Pisolitic** concretions and internal reworking are very common in the crust. The casings. almost vertical, often aligned along root conduits, may mark the base of the crust.

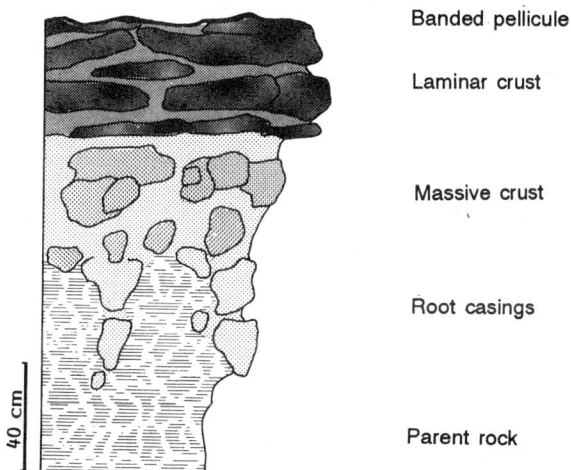

Fig. 11. *Examples of evolved* **calcretes.** *Similar calcretes are presently found in many regions with Mediterranean or subarid climate; their depth is highly variable. They may have possibly been covered by a more or less rubefied pedological horizon.*

— *Micromorphology*: Calcretes often consist of relatively homogeneous micrites, totally recrystallised locally. Banded and pisolitic facies define a characteristic contrast (Fig. 12a and b).
— *Environment of formation*: The zones of **phreatic water circulation** may be adequate for the formation of calcretes but a *pedological environment* is certainly the most effective and gives rise to fully evolved profiles (Fig. 11). The extent of the contribution of biological activity in the fixation of limestone is now fully established. For instance, fungal remains were found in semi-arid calcretes (Verrecchia, 1990, 1991).

Calcrete may develop *at the base of a pedological profile* but depending on the protection of a plant cover, it can also develop *at the surface* where only micro-organisms ultimately intervene; thus banded pellicule may be formed directly on exposed limestone due to the activity of micro-organisms (Krumbein and Giele, 1979), but it may also form under an unconsolidated horizon covered with vegetation (Calvert and Julia, 1983; Julia and Calvert, 1983).

— *Mechanism*: This may be simple **cementation** by calcite crystals invading pores and fissures; this cementation probably displaces the existing elements, creating a disturbance

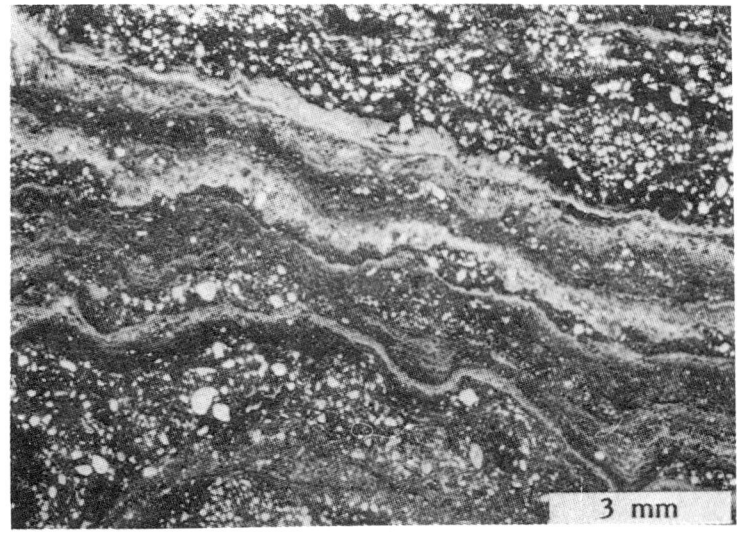

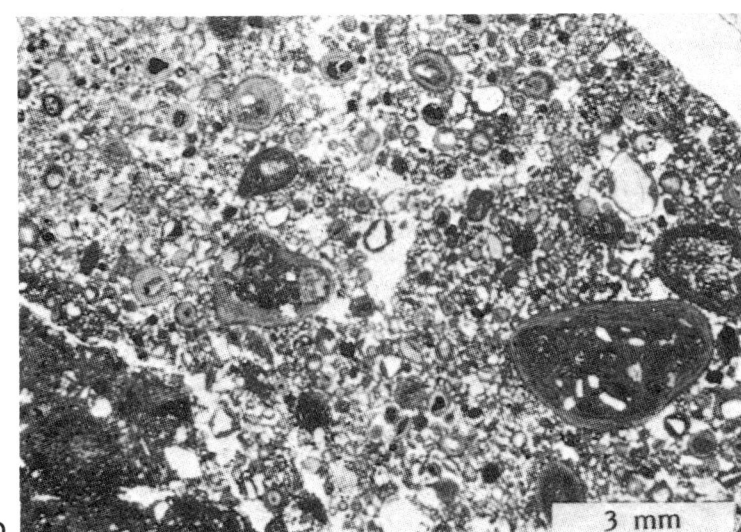

Fig. 12. *Micromorphological characters of recent* **calcretes.**

a) Banded pellicule at the top of calcrete (central Algeria). The irregularity of fine laminae of carbonates indicates a biological origin. Microscopic view in natural light.

b) Mineral and lithic elements surrounded by laminated limestones which give them a pisolitic appearance. Calcretes on basaltic rocks in the Canary Islands. Microscopic view in natural light.

in the facies (Kulke, 1974; Watts, 1978). The most important mechanism is **epigenesis** (Millot et al., 1977; Bech et al., 1980) or the replacement of minerals in the parent rock by calcite (Paquet and Ruellan, 1993). It is the major phenomenon of surficial evolution in **arid zones** (Halitim, Robert and Pedro, 1983). Recent studies are oriented towards research on early metastable minerals, which may be aragonite (Nahon et al., 1980) or *whiskers of slight magnesium-bearing calcite* (Calvet and Julia, 1983). Degradation of argillaceous minerals at the top of the profile is the key relation between silicates and carbonates in these crusts (Paquet, 1983).

2.3.3.2 Fossilised calcretes in the Thanetian of Champagne

In this region Senonian (Cretaceous) *chalk* is systematically overlain by a weathered level that varies in thickness from 2 m to 5 m. The pure chalk, without flint, changes towards the top into a matrix of angular hardened yellow chalk grains (Fig. 13). The chalky matrix grades from white to grey, becomes **granular**, then **nodular** and hardened. Towards the top the alterite is very hard and presents a foliated structure, marked by broad horizontal, irregular fissures. The profile is topped by **sands with microcodiums,** which rework the elements of chalk and alterites equally. Microscopic studies have shown that the chalk was initially micrite, then recrystallised into irregular rhombohedral crystals (Fig. 14a); live microcodiums traverse the top of the alterite (Fig. 14b). Geochemical evolution indicates heavy weathering of alumino-silicates; an argillaceous assemblage evolves (Fig. 13); interstratified **kaolinite-smectite** signifies the disappearance of kaolinite in favour of smectite.

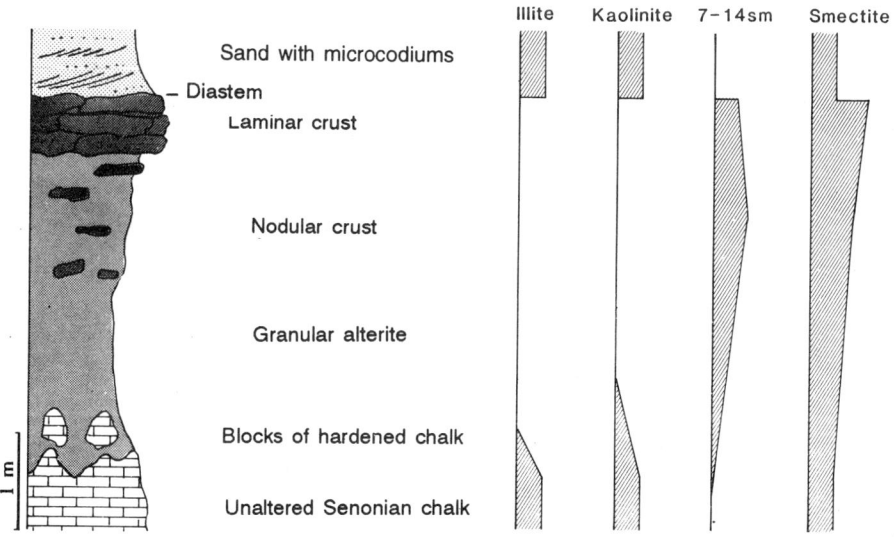

Fig. 13. **Calcrete** *at the top of chalk in the Reims Mountain. This calcrete, systematically present in this region, is sealed under Sparnacian sediments. The evolutionary trend of the clay minerals is only of semi-quantitative value.*

This alterite at the summit of the chalk dates back to the end of a long phase of emergence, which signifies the transition from Cretaceous to Tertiary in the Champagne region (Laurain and Meyer, 1979). The following reasons favour its interpretation as calcrete:

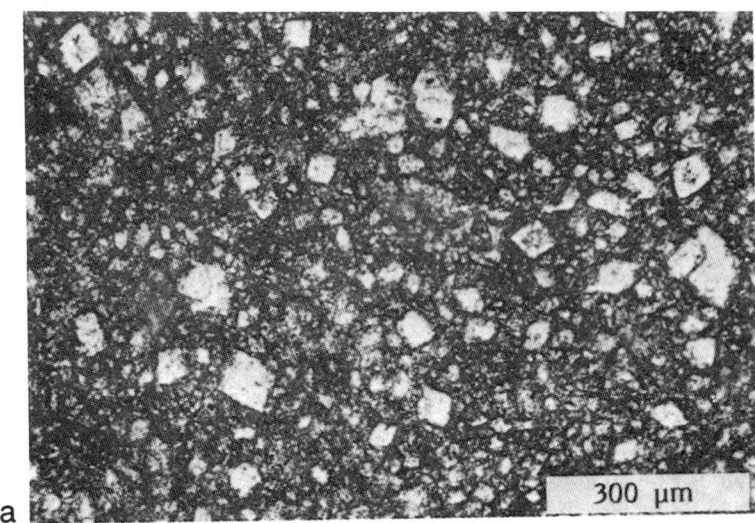

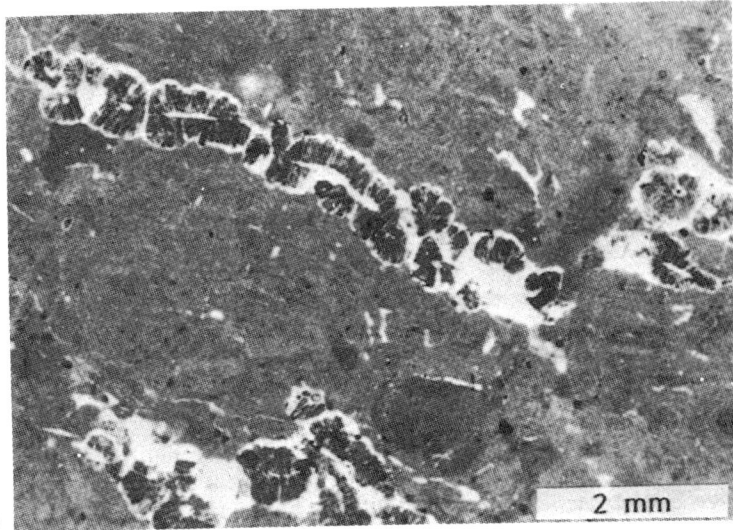

Fig. 14. *Micromorphological characters of Thanetian* **calcrete** *in Champagne.*

a) Recrystallisation of chalk into heterometric rhombohedral crystals of calcite. Microscopic view in natural light.

b) Destruction of chalk by **microcodiums**; *the corn-ear forms show cells made up of monocrystals of calcite. Microscopic view in natural light.*

— The field profile is very similar to the recent calcretes developed on chalky substratum, for example in Israel (Yaalon, 1978).

— Classical recrystallisation of calcite in calcrete as represented by large rhombohedral crystals (Folk, 1971).

— The microcodiums found from the Cretaceous to the Pleistocene signify phases of emergence and, more often, of paleosols (Bodergat, 1974; Freytet and Plaziat, 1982, p. 42). They probably imply seasonal contrasted climate with a marked dry season (Bignot, 1995).

2.3.3.3 Calcretes in ancient sequences

A large number of geological reports describe calcretes of various ages. Excluding those descriptions too brief to serve as references, a bibliography may be proposed for the more important ones.

Calcretes formed during the *Cenozoic* are distinguished by the fact that they formed on stable cratonic regions and are presently outcrops (e.g. in the Sahara: Conrad, 1969), or they were fossilised in the midst of continental deposits. The latter certainly proves to be the best choice for interpretation, since they are no longer active. Some excellent examples have been reported by Freytet and Plaziat (1982), Crouzel and Meyer (1975), Thiry (1981), Truc (1975), and Valleron (1981).

Various types of calcretes belong to the *Mesozoic* up to the end of the Cretaceous (Freytet, 1971). Continental Jurassic deposits may also be rich in pedological calcretes (example, west coast of Portugal at Cape Mondego, near Figueira da Foz). However, the maximum number of examples are from the Triassic, particularly in sandstone facies (Durand and Meyer, 1982; Esteban et al., 1977; Marzo, Esteban and Pomar, 1974; Ortlam, 1971).

Calcretes have been described from the Permian in the *Paleozoic* (Röper and Rothe, 1975; Durand and Meyer, 1982) and are occasionally related to mineralisation (Barbier, 1978). In the Carboniferous of Wales the carbonated accumulations are sufficiently varied to be interpreted as true calcrete (Wright, 1982b) or as fossil **rendzina** (Wright, 1983). The Siluro-Devonian **Old Red** Sandstone of Great Britain encloses several layers of calcretes (Allen, 1974; Watts, 1977); an example is given in Fig. 15. These authors emphasise the disturbance connected with the growth of crystals, resulting in the appearance of *pseudo-anticlines* in the crust.

Regarding the *Precambrian,* though many paleosurfaces and paleosols have been identified (Retallack, Grandstaff and Kimberley, 1984), calcretes have been recognised only as an exception (Bertrand-Sarfati and Moussine-Pouchkine, 1983). This may be due to the conditions prevailing during that epoch, or simply due to the *solubility of carbonates* which could have been remobilised during the entire Phanerozoic era. The relevance of paleosols to Precambrian atmospheric composition has been discussed by Retallack (1989).

2.3.3.4 Polyphased calcretes

Carbonated accumulations leading to calcrete may have been the consequence of several processes:

— Formation of concretion within the limits of **fluctuations in the phreatic water table.**
— Deposition of cementing material in the porous facies, e.g. the *base of detrital* **sequences.**
— Formation of *pedological concretions* which evolve in two contrasting fashions depending on whether the landscape undergoes **erosion** or **accretion;** in the former calcrete develops at the base and degrades towards the surface, while in the latter it develops at the *summit,* with complete assimilation of the surface soil horizons.

These processes could have occurred simultaneously or in succession; diagenesis may have in addition remobilised the limestone and caused an almost ascending migration. All

Fig. 15. *Dolomitic crust in the Silurian Old Red Sandstone. Accumulation of carbonate extends downwards through tubules which are more or less perpendicular to the horizontal milieu of genesis. Lydney, Gloucester, Great Britain (sample presented to the author by J.R.L. Allen).*

these resulted in the formation of complex calcretes, which can be interpreted only after deciphering the successive phases of evolution. Schematic examples of these polyphased evolutions are depicted in Fig. 16, as found in the fossilised Oligo-Miocene molasse of central Aquitaine. A similar morphology is found in many continental deposits.

2.3.3.5 Fossilised dolocretes in the Permian of Saint-Die Basin (north-eastern France)
In this basin the thickness of Permian sedimentary and volcano-sedimentary deposits may exceed 250 m. The Saint-Die beds, probably of Thuringian age, are nearly 100 m thick. **Dolomite** is commonly present as small deformed concretions or as volutes, which are

---

Fig. 16. *Successive phases of carbonate accumulation in the Oligo-Miocene molasse in Aquitaine. Each figure represents about 1 m depth.*

a_1. **accumulations** *related to fluctuations in the water table. This accumulation is shown in Figs. a_2 and a_3.*

b_1: *an increase in* **permeability** *of the sediment in the channel causes formation of concretions along bedding foliations; this accumulation becomes a true crust in Fig. b_2; the crust degrades as shown in the upper part of Fig. b_3 due to lateral circulation.*

c_1: *formation of concretions along root conduits; these conduits coalesce, giving rise to a true crust under the rubefied B_t horizon, as shown in Fig. c_2; in Fig. c_3 the crust at the summit is degraded, indicating a change in hydraulic regime.*

d_1: *pedological crust on a rubefied horizon, as in Figs. d_2 and d_3; successive addition of sediments in small quantities led to crust of considerable thickness.*

EXAMPLES OF PALEOALTERITES AND PALEOSOLS

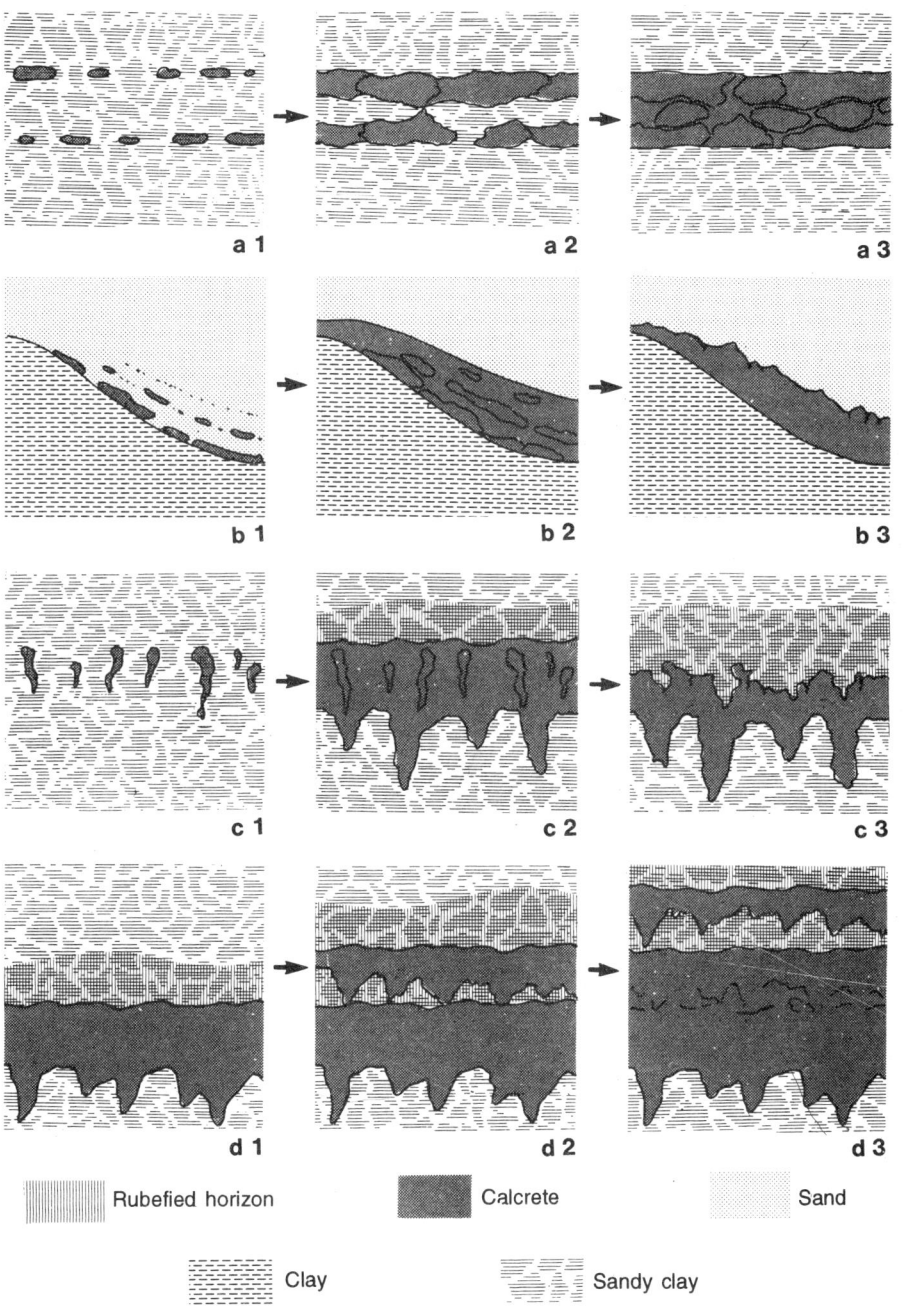

Rubefied horizon   Calcrete   Sand

Clay   Sandy clay

associated with decolorisation. Its distribution appears to be random on the outside of three consistent encrusted layers. The following example refers to the upper layer (Meyer, 1981).

The dolomite crust is **massive, coarsely foliated,** and extended downwards by *excrescences that are roughly cylindrical* but occasionally *dichotomous* (Fig. 17). The dolomite is creamy-white, the crystals often visible to the naked eye, practically non-ferriferous, inhomogeneous, even though detrital elements are rare or absent, with a discernible nodular form. The surrounding feldspathic sandstone is coarse grained, poorly graded and poorly stratified. A deep red argillaceous matrix imparts a feeble cohesion to it. The gradation of sandstone to dolomite is generally characterised by a greenish clay accumulation, a few centimetres or millimetres in thickness. The clay fraction of any colour is a well-crystallised illite. The characteristic microfacies at the top of the crust are depicted in Fig. 18.

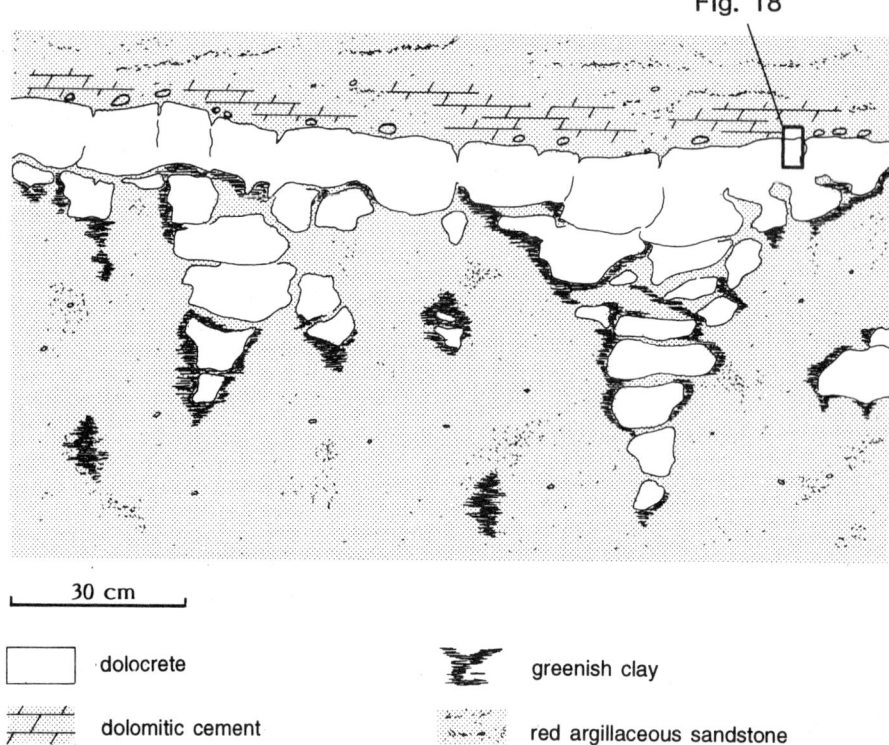

Fig. 17. **Dolcrete** *in the Permian of Saint-Die Basin. Dolomitic cement (facies C in Fig. 18) is a late deposit consequent to relative impermeability of dolocrete.*

The essential features of the outcrop suggest a crust of pedological type: the coarse foliation and almost vertical and cylindrical concretions are comparable to those of present-day crust (see Fig. 11). The internal heterogeneity of the crust also supports a similar interpretation; shapes and blobs which are related to **nodules and fine laminae** at the top of the crust probably resulted from the recrystallisation of **banded pellicule.** The dolomite

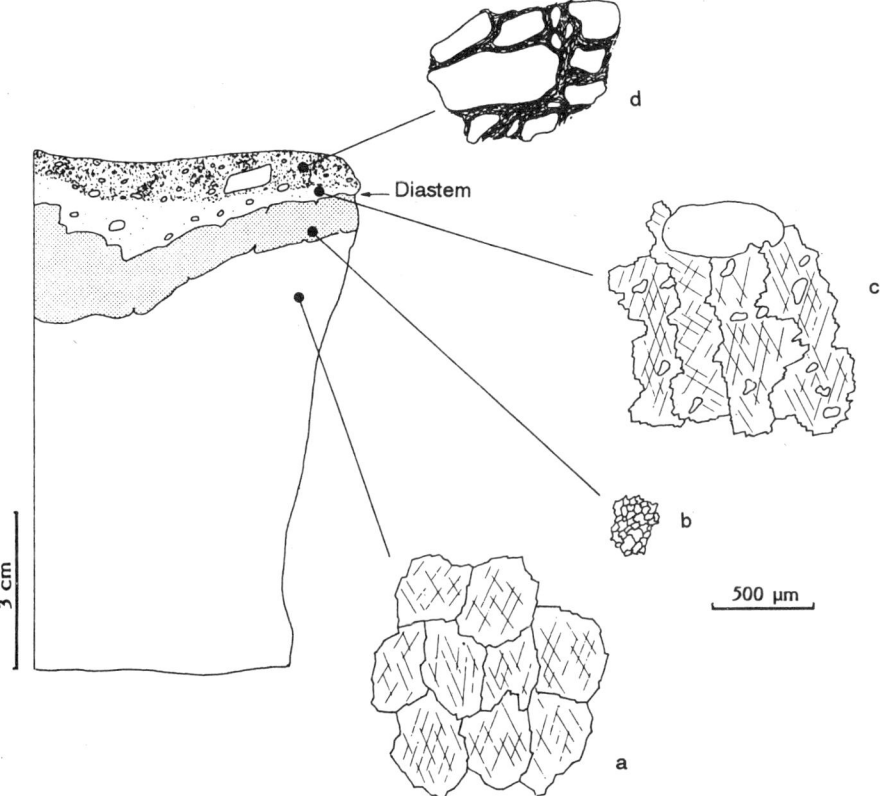

Fig. 18. *Detailed view of the top of the* **dolocrete** *represented in Fig. 17.*

a) Large crystals of dolosparite in polarised light; a crystal has a rolling or undulatory extinction.
b) Fine laminae of microdolosparite.
c) Poikilitic cyrstals of dolomite, elongated vertically, and partly enclosing corroded quartz crystals.
d) Detrital grains (quartz, feldspars) in an argillo-arenaceous matrix.

crust is very poor in residual minerals: **epigenesis** of silicates by carbonates must have played an important role, as has been established in recent calcretes (Millot et al., 1977).

When one considers that the formation was buried to a depth of 2000 m, an important question arises: whether the dolomite developed in the environment of sedimentation or resulted from diagenetic dolomitisation of a calcrete. Homogeneity of the argillaceous assemblage proves an evolution related to burial and large dolomite crystals indicate **diagenetic** dolomitisation which, however, could have been at a relatively early stage since there is almost no iron in this dolomite (Richter and Füchtbauer, 1978).

2.3.3.6 Fossil dolocretes: Interpretations

The problem of fossil dolocretes as envisaged in the previous section occurring in a number of sequences, presents the choice between a *surficial origin* or a *diagenetic origin* for dolomite.

Dolomitisation in present-day soil is rare but not unknown. An example in calcareous soils has been reported by Sherman, Schultz and Alway (1962). It is in fact dolomitisation replacing limestone at the base of the profile in contact with the fluctuating water table. The same authors have also reported authigenic dolomites in **solonetz**. Though recent crusts enriched in magnesium are rare, nevertheless they exist in various regions all over the world (Goudie, 1973, p. 21). Watts (1980) observed imposing Quaternary calcretes in Kalahari of which some are highly dolomitised. Nahon (1976), working on carbonated crust of Senegal, observed precipitation of dolomite at the base of the profile in the downstream zone. This precipitation might be related to the destabilisation of palygorskite and consequent liberation of magnesium at the summit of the profile (Millot et al., 1977; Paquet, 1983).

Though fossil dolocretes appear to have been formed due to diagenesis during burial, the above-mentioned examples show that a surficial origin cannot be completely ruled out. A few examples are cited here which favour this interpretation:

— Dolomite at the base of calcrete in the Stampian of Central Aquitaine (Crouzel and Meyer, 1977b); it is a dolomicrite in a sequence that has undergone no marine influence or any evidence of burial.

— Dolomite at the base of several calcretes of the Eocene in Tunisia (Sassi et al., 1984); repetition of the phenomenon excludes dolomitisation related to late circulation.

— Dolomite of **violet zones** of the Vosges Buntsandstein (Durand and Meyer, 1984); electron microprobe studies revealed inclusions of dolomite in the early silicified facies.

2.3.3.7 Accumulation of siderite

Siderite ($FeCO_3$) is known to precipitate in continental waters or during diagenesis influenced by meteoric water (Matsumoto and Iijima, 1975; Richter and Füchtbauer, 1978) when the environment is reducing and the S^{2-} ions have low activity; in fact the abundance of sulfur results in *sulfides*.

The accumulation of siderite may also indicate **paleosurfaces** (Perel'man, 1967, p. 193) but is independent of crusts, mentioned earlier, which appear in an oxidising environment. Contrarily, it indicates the presence of *freshwater marshes* in which a fixation of iron occurs, which is generally derived from the source area as organo-mineral complexes fixed on clay particles.

2.3.3.8 Problems of detection and principles of interpretation

Fossil carbonated crusts can be identified relatively well by comparison with many present-day models (Sec. 2.3.3.1). Certain important factors should be kept in mind by the investigator:

— **Banded pellicules** and **pisolites** of calcretes may have a structure identical to that of **stromatolites** or **oncolites** forming in an aquatic environment.

— After fossilisation, calcareous mud deposited in a **palustrine** environment may acquire a facies close to **calcrete,** the difference between the two being ascertainable through paleoecological criteria (Plaziat, 1984).

— The *overall morphology* of the carbonated crust at the metre or decimetre scale is generally better preserved than the **microstructures**, which alter due to diagenetic recrystallisation.

— In the fossil carbonated crust the *mineralogy* of the carbonates varies, for example magnesium or ferriferous calcite, ankerite, siderite; the mineralogy must be studied as a function of the environment of the formation considered together with **diagenesis**.

— **Microcodiums**, when they exist, are very characteristic of paleosurfaces; they may invade calcareous or marly layers, which *excludes dolomite* (P. Anadon, pers. comm.).

Study of fossil crusts justifies two modes of interpretation, the **paleoclimate** and the **duration of breaks** in sedimentation:

— Calcretes, when well developed, indicate *warm and relatively dry climates*; the presence of palygorskite and dolomite indicates a distinct water deficit.

— Calcretes of pedological origin assume the form of paleosurfaces, which generally suggests breaks in sedimentation longer than for calcretes associated with groundwater (Fig. 16).

— The duration of formation of calcrete or dolocrete is relatively less at the *lower level of the landscape and increases towards the higher level*.

— The duration of formation of calcrete is relatively shorter on a calcareous parent rock than on a silicate parent rock containing calcium. The duration is subject to the hazards of external contributions when the substratum is devoid of calcium.

— The duration of formation of calcrete increases if the porosity of the substratum is poor.

CALCRETES

Characteristic morphology:
Lamellar crust very hard on the surface; root casings; white to salmon colour.

Characteristic facies:
Fine laminations, onchoids.

Microfacies:
Mainly micritic, occasionally rhombohedral microsparite.

Mineralogy:
Calcite, sometimes dolomite.

Microcodiums:
Often present in the Mediterranean basin, from the Cretaceous to the Pleistocene.

Parent rock:
Carbonated (example molasse) or calcitic (example basalt)

Substrata:
May be very diverse (granite, sandstone etc.)

Frequency:
Very frequent, known from the Precambrian to the present day.

Climate:
Mediterranean to desert.

Convergence:
Lacustrine limestone, generally more homogeneous.

2.4 Sulfide or Sulfate Paleoalterites

2.4.1 Present-day Sulfide and Sulfate Soils

Gypseous crusts (**gypcretes**) which occur in a desert environment, for example in southern Tunisia, are generally related to the presence of pre-existing gypsum (older rock or aeolian), which dissolves and reprecipitates at close proximity to the soil surface or within a few metres from the surface. Crystals are visible to the naked eye and the assemblage has no specific structure (Durand, 1959). These gypcretes are often associated with *gypsum rosettes*, which form by the action of feebly saturated solutions above the water table in a non-gypseous porous sandy sediment (Pouget, 1968).

With the exception of the previous example, soil containing sulfides and sulfates are called **acid sulfate soils;** they have been classified as saline soils because the role of the Na^+ ion is generally of considerable importance (Duchaufour, 1977, p. 453). It usually evolves in two phases: **reduction phase** which leads to the formation of sulfides, followed by an **oxidation phase** which produces sulfates.

These particular soils develop in the vicinity of an evaporite or marine environment and their mineralogical evolutions are similar. Fig. 19 gives an idea of such classical reactions.

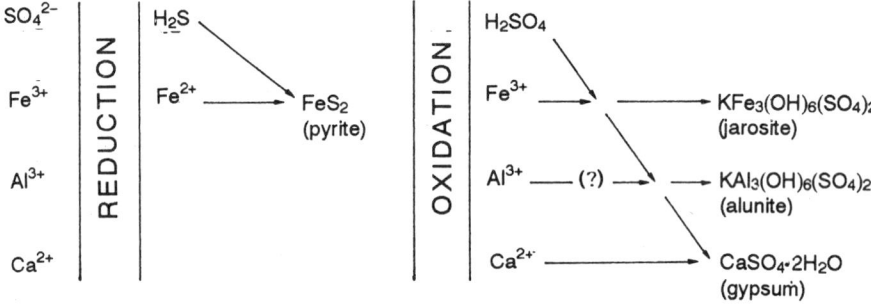

Fig. 19. *Schematic diagram for mineralogical transformation susceptible to ingression in acid sulfate soils. Ingression of micro-organisms regulates these transformations. Though the formation of alunite is thermodynamically possible under these conditions, its presence in the acid sulfate soils has not been clearly established.*

These soils acquire their supply of **sulfur** by several modes:
— **leaching** *of sulfate rocks or older sulfide rocks;*
— *abundance of* **organic matter** *in the parent rock;*
— *sea-water.*

Many acid sulfate soils are associated with **coastal environments,** including polders as well as **mangroves,** since the climate is sufficiently warm (Pons, 1972; Van Breemen, 1972). The type of sulfate which is susceptible to precipitation depends on the general environment. Baltzer (1970), for example, described the mangroves of New Caledonia where only **gypsum** is precipitated, while Vieillefon (1974) observed **jarosite** in the Casamance mangroves. The latter author proposed a model for the evolution of mangroves in estuaries during the course of accretion (Fig. 20). Pedological differentiation affects a

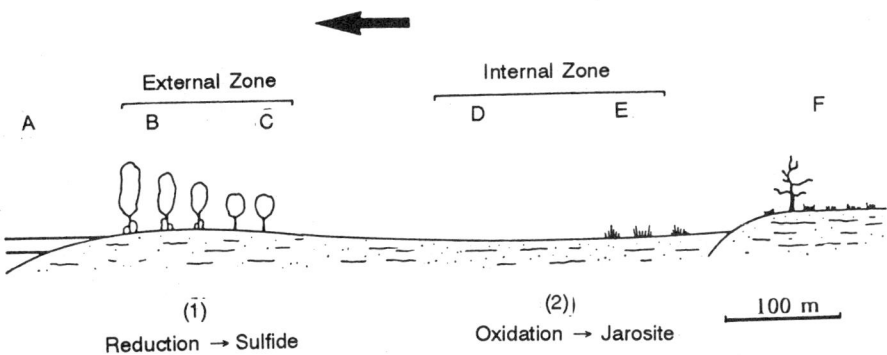

Fig. 20. *Theoretically established succession of environments in present-day tropical* **mangroves** *(Casamance estuary, Senegal, after Vieillefon, 1974).*

   A. *Estuary: water level varies with tides and seasons.*
   B. *Mangrove with Rhizophora.*
   C. *Mangrove with Avicennia.*
   D. *Bright spots, extra saline, without vegetation.*
   E. *Halophyte spots, commencement of freshwater influence.*
   F. *Ancient terrace.*

Soil (1) appears in environment B; it is subsequently totally reduced. Arrow indicates the direction of changing environment over time. This change is due to alluvial accretion. Soil (1) evolves gradually to soil (2), observed in D and E, which is oxidised, rich in jarosite and compact at the surface.

thickness of 60 to 80 cm; the succession of characteristic facies is as follows, from the top downwards:
— *Oxidised grey-brown* surficial horizon with rust patches rich in alkaline earth ions with no trace of sulfide components;
— *yellow horizon with jarosite,* very plastic;
— *clear grey horizon,* with no pyrite or jarosite;
— profoundly *reduced* horizons with pyrite and organic matter.

It is necessary to note the high potentiality of certain soils to initiate the neoformation of varied sulfates according to their habitus and mineralogy: **anhydrite, barite, celestite** have only been cited because of their greater probability to fossilise (Hanna and Stoops, 1976; Stoops, Eswaran and Abtahi, 1977).

2.4.2 Paleosols with Gypsum and Anhydrite

Sulfates are evaporite minerals, which implies that they have greater solubility compared to many other minerals. Therefore, *sulfated pedological accumulations, of limited volume,* have a high probability for *going into solution, recrystallisation* or fixation elsewhere during fossilisation of soil. Thus it is difficult to suggest a characteristic facies for fossil gypcrete, which can often only be identified by associated facies (root zones, calcretes or aeolian dunes), even by **boxworks**, or by the pseudomorphs of crystals of gypsum and anhydrite.

2.4.2.1 Fossil gypsum rosettes in the Ludian of Paris Basin

In the Paris Basin, east boundary, in the Ludian (Eocene), bands of dolomite enclose gypsum rosettes, either as boxworks or as carbonated or silicified pseudomorphs. Inter-

vening mineralogical transformations have been established through petrological studies (Meyer, 1981).
- **neoformation** of large gypsum crystals;
- **replacement** of gypsum by ferriferous calcite;
- **decomposition** of ferriferous calcite leading to pure calcite and goethite;
- selective **silicification** of gypsum rosettes.

As this formation has undergone no burial, its entire evolution took place in a surficial environment, wherein the geochemical parameters varied considerably. Replacement of gypsum by ferriferous calcite proves that **sulfur** leaves the environment very early and that ferrous ion goes into carbonates and not into sulfides.

Interpretation of this evolution seems possible: Gypsum rosettes appear in a **vadose medium,** they signify the emergence of zones from basic phreatic surfaces, are capable of inducing precipitation of a complex mineralogical suite and the climate tends to be **arid.**

2.4.2.2 Lenticular gypsum and anhydrite crystals

The small isolated or twinned crystals of lenticular gypsum have a habitat similar to that of the gypsum rosettes, and statistically appear in the vadose zone (Bertrand and Jelisejeff, 1974). This gypsum is not a pedological index, but rather an index of emergence defining the *hydraulic regime*. Such crystalline forms are of interest since they fossilise easily as boxworks or pseudomorphs, identifiable in older sequences (Fig. 21a).

It has been established that crystals of gypsum may thrust against the enclosing grains during the process of their growth (Plet-Lajoux, Monnier and Pédro, 1971). This is one reason for mechanical degradation, by *opening up fractures* and *splitting* rocks which outcrop in desert environments (Cortelezzi and Kilmurray, 1965; Yaalon, 1970). The phenomenon does not allow for large quantities of gypsum but sometimes causes spectacular disorders; gypsum crystals readily dissolve and fissures may be filled with aeolian sands or calcite deposits (Laville, pers. comm.).

Anhydrite crystals, which *crystallise early in certain desert environments* may fossilise after epigenesis. They therefore act as precise indexes facilitating identification of these environments (Fig. 21b).

2.4.3 Paleosols with Jarosite in the Lignites of Soissonnais

In an open-cast mine at Verzenay (Marne) many **root traces** were found on top of the Soissonnais lignite and **differentiation** into horizons is clearly recognisable in the field (Gruas-Cavagnetto, Laurain and Meyer, 1980a): A paleosol comprising four readily recognisable horizons is described below from the top down (Fig. 22):

- A *grey horizon* with lignite remnants, less hardened, porous, at places nuciform.

- A *yellow earthy* massive horizon, hard, of polyhedral structure, the faces with **slickensides;** a number of root traces, sometimes marked by a flaky yellow deposit, analysed as pure **jarosite** by the X-ray diffraction method, appear as granular aggregates, nearly opaque in natural light and opalescent in polarised light; distributed homogeneously throughout the horizon.

- A clear grey horizon, silty-clay, made of particles, traversed by root conduits; quartz and muscovite occur in a heterogeneous matrix where plasmic differentiation of **skelsepic** and/or **lattisepic** types are seen.

- A black horizon with lignite, at places as small flakes.

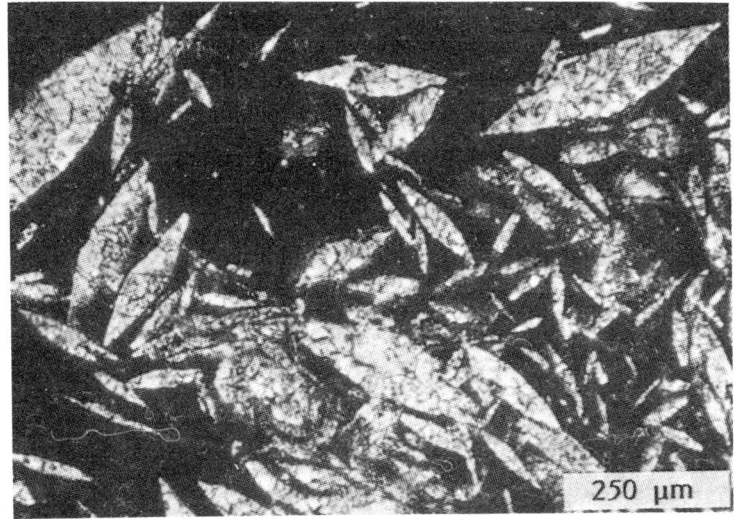

Fig. 21. *Pseudomorphs of* **sulfate** *crystals.*
 a) Lenticular crystals of **gypsum** *developed in micritic matrix; they were pseudomorphed by sparite aggregates. Miocene in Aquitaine. Microscopic view in natural light.*
 b) Silicified crystal from the violet zone of Buntsandstein in southern Vosges (France). The crystal form and traces of cleavage parallel to the faces suggest **anhydrite;** *silicification is early in the facies; hence anhydrite is an indicator of the environment. Sample from M. Durand.*

The mineral assemblage of clay is dominated by kaolinite, then illite and smectite, the percentage of which shows no significant change. **Palynological studies** have shown that the flora in the profile is very diverse (Gruas-Cavagnetto, Laurain and Meyer, 1980b), and

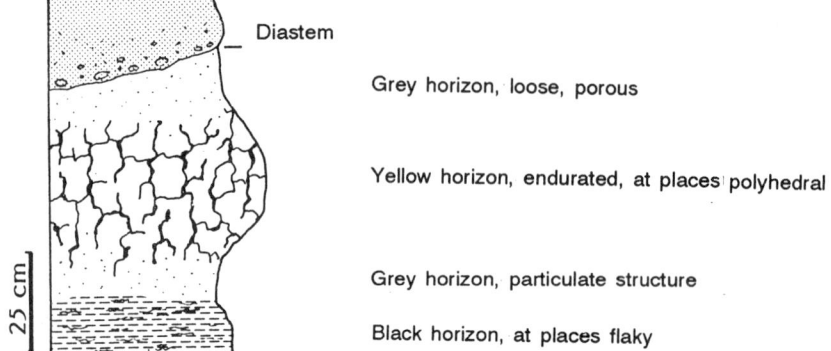

Fig. 22. *Acid sulfate paleosol in lignites of Soissonnais. Differentiation in four horizons, each with specific structure, is shown (Verzenay, Marne).*

is *marine influenced* towards the base, which gives way to a vegetation poor in species, and of *continental tendency in the surficial horizons*. Lower horizons include three known forms living in a present-day mangrove environment (*Nypa, Bruguiera*, and spores similar to those of *Acrostichum aureum*).

Jarosite is often described as a recent alteration product of sulfides (Routhier, 1963, p. 244). However, it is quite possible that impregnation of jarosite in the yellow horizon dates the environment of the deposit.

— Gypsum is the sulfate which is the product of weathering, the necessary calcium having been derived from carbonated Upper Eocene deposits.

— Impregnation of jarosite throughout the horizon is uniform and does not conform to epigenesis by the circulation of waters, which preferentially follow fissures.

— The jarosite horizon has developed laterally and is independent of the diastem at the top of the profile as well as of the existing topography; this could not have been possible had it been a late formation.

The profile can be interpreted as an **ancient acid sulfate soil**. The environment can be determined by comparison with the observations made in the inner zones of the mangroves of Casamance (Vieillefon, 1974). Palynology confirms that the soil belongs to the **inner zone of the mangrove** if the deficiency of a pollen-bearing assemblage in the surface horizon is considered; only the **later pollen** survived the oxidising conditions which were prevalent at the surface and were incorporated in these soils. This paleosol corresponds to the soil of type 2 of the landscape schematised in Fig. 20.

2.4.4 Jarosite, Alunite and Barite in Paleoalterites

2.4.4.1 *Jarosite*

Jarosite has long been known to form by alteration at the top of sulfide sediments (iron hat). The preceding example shows that it may also indicate the presence of an acid sulfate paleosol; it would be prudent, however, not to interpret it as such *except in combination with other indexes*. Plaziat (1975) also proposed this mineral as an index for ancient mangroves, even though he preferred paleontological criteria, such as mangrove oysters, which have a characteristic morphology (Plaziat, 1970).

Guendon (1981) reconstructed the paleogeography of the Cretaceous of Vaucluse on the basis of jarosite-bearing acid sulfate soils that had developed on grey argillite by oxidation of **pyrite**. Natro-jarosite is formed by replacement of the K^+ ion by the Na^+ ion in a medium rich in sodium. A specific environment is indicated when the **pH** value alternates between alkaline and acid (Perinet, Taieb and Tiercelin, 1980; Goldbery, 1978).

2.4.4.2 Alunite

Even though alunite has a crystal structure similar to that of jarosite, a solid solution between the two minerals hardly ever occurs in nature (Brophy, Scott and Snellgrove, 1962). This indicates unique conditions of formation for the two minerals. Though the **hydrothermal** *origin* of alunite is recognised, its presence in acid sulfate soils can only be assumed, not established (Van Breeman, 1972). However, strong geological evidence does establish that the formations enclosing alunite appear in surficial continental environments, away from all hydrothermal activity (Kashkaj, 1961). The following examples furnish the evidence:

— In the opal beds associated with silcretes of southern Australia (Coober Pedy) alunite is common. It occurs disseminated in the rock or in clusters of a size measurable in metres; it precedes the deposition of opal (Jones and Segnit, 1966).

— On the western border of Portugal (Meyer and Pena dos Reis, 1985), a silcrete (Oligocene or Lower Miocene) several metres thick has fissures at the top filled with alunite (Fig. 23b) which has migrated from the underlying formations. This example is described in detail in the section dealing with silicification (Sec. 2.5.2).

— In the Paleogene of the Duero Basin of Spain *silicification affected each cyclothem* after its deposition (Blanco and Cantano, 1983). The phenomenon is manifested as an opal deposit, a neoformation of smectite and alunite. Alunite occurs sporadically but locally lenses of it may attain a thickness of 50 cm.

— In the Jurassic argillite of Israel (flint clays), Goldbery (1978) demonstrated evidence of beds of *natro-alunite* concretions. He attributes this mineral to a diagenetic transformation of pyrite disseminated in the associated carbonaceous facies.

A few remarks in conclusion: *Alunite is often associated with silicified facies* and could have had its origin in an *aggressive environment* where many alumino-silicates are unstable. The macroscopic colour is creamy-white, which makes it difficult to identify since it is intimately mixed with siliceous facies, which leads one to question whether it actually occurs more frequently than thought. Alunite could be used for dating by the K/Ar method (Blanco and Cantano, 1983) but great care must be taken in applying this method. Such sulfates may crystallise repeatedly, causing the geochronological watch to return to zero each time. Dating of alunite in the beds of Tolfa (Italy) associated with trachyte provided a recent age (M. Nicoletti, pers. comm.).

2.4.4.3 Barite

Neoformed barite occurs in certain present-day soils (Stoops, Eswaran and Abtahi, 1977). It is found in some Tertiary siliceous crusts (Fig. 23c), where its origin is surficial (Meyer and Pena dos Reis, 1985). This shows that an affinity exists between SO_4^{2-} ions and Ba^{2+} ions. Such an affinity explains the existence of barite at diverse stages within a geological cycle. Barite concretions at discordant surfaces, for example between the Hettangian and Hercynian basement in Perigord (Daugas, 1981), appear to be **contemporaneous to burial diagenesis,** whereas those observed by Davaine (1980) in Morvan are the product of early mineralisation. Barite should not always be attributed a pedological origin, but only when definite evidence is available. Moreover, the phenomenon of meteoric weathering some-

Fig. 23. **Sulfate** *crystals of supergene origin.*

a) Rhombohedral pseudocubics of **jarosite** *in mangrove paleosol. Sparnacian of Champagne; observation under SEM.*

b) Rhombohedral pseudocubics of **alunite** *intimately admixed with lepispheres of silicates in silcrete from Portugal, Tertiary. Observation under SEM.*

c) **Barite** *crystals in the same silcrete from Portugal, Tertiary. Observation under SEM.*

times could result in simple *redistribution*: Parron, Nahon and Tardy (1980) interpreted the contemporary baritic bed of Chaillac as an accumulation at the base of lateritic ferruginous cuirasse, overlying a stratified bed.

2.4.4.4 Diagnosis of fossil mangroves

The first point to be considered is the context of geodynamics. Mangroves often grow on *alluvial deposits,* which induce *accretion of the emerged areas.* If for some reason the **regressive** tendency gives way to a **transgressive** tendency, the phase of sulfate reduction is not followed by the phase of oxidation: the mud is buried in a reduced condition and the sediment reflects no characteristic mineralogical evolution (Lucas, Kalk and Gouleau, 1979); subsequently a normal evolution of marine sediments follows, even if the sulfide percentage is relatively significant.

The second point concerns stratigraphy: to the best of our knowledge, pre-Tertiary mangroves, even if present, have not been reported (Plaziat, Koenigen and Baltzer, 1983).

A diagnosis based on paleontological evidence, such as **oysters** or **pollen**, is definitely the best when such remnants are present (Plaziat, 1975).

A diagnosis based on the presence of **sulfate** is possible but must be robustly supported by the recognition of real profiles, to avoid confusion with recent sulfation or with sulfates which are not in the environment of their genesis but of late migration.

2.4.5 Principles of Interpretation

Crusts rich in gypsum, even in anhydrite, indicate arid climates. Although these minerals are very mobile, there is a good chance of finding traces of crystalline forms as **boxworks** or as products of **epigenesis,** in particular siliceous.

SULFATE HORIZON

Mineralogy and facies
Gypsum   : twin lenticular crystals (gypsum rosette) in vadose zone.
Jarosite : powdery concretions, yellow flakes.
Alunite   : massive concretions, microcrystalline, very often whitish-grey.
Barite   : concretions of lamellar crystals, often pink; labile (unstable) mineral often found as boxworks.

Substratum
— parent rock argillo-sulfide or lignitus for jarosite and alunite;
— porous sand for concretions of gypsum;
— silicified layers for alunite.

Frequency
In all epochs; alunite beds have a cyclic stratigraphic distribution more or less following that of bauxites.

Paleoenvironment
Desert for gypsum and anhydrite; acid sulfate soils, near the sea, for gypsum and jarosite, forming from mangrove soils when the climate is warm; often diagenetic for barite.

Convergence
Oxidation of sulfide is not always linked to the environment of deposition; it may be recent (iron hat).

Acid sulfate paleosols may occur at the shores of lakes which have a specific chemistry, or at the periphery of marshes if an external supply of sulfur exists. In the vast majority of cases however, these paleosols indicate a littoral environment influenced by sea-water. Hence they are good indicators of *paleogeography* and of *paleoenvironment of an aggressive changing chemistry,* which might have played a major role in the genesis of mineral concentrations (Perel'man, 1967).

Acid sulfate paleosols do not systematically correspond to environments of mangroves. This implies a *specific vegetation* which grows only in climates more or less humid but *always warm* (absence of frost).

2.5 Siliceous Paleoalterites

2.5.1 Facts and Models

2.5.1.1 Silica in continental environment
Silica has a very complex chemistry which has been exhaustively analysed by Iler (1979). Continental waters *invariably carry silicon (Si) in solution*: 5 to 30 ppm in subsurface water and 15 to 20 ppm in soils (Wilding, Smeck and Drees, 1977). During meteoric weathering of rocks the liberated silica may be transported as solution in natural waters. Part of the silica is carried to the ocean but part remains on the continent, most often very close to the place of its liberation.

In present-day continental environments silica precipitates in various mineralogical forms which merit precise schematisation. Mitchell (1975) proposed the following schematic model: monomeric silicic acid H_4SiO_4 may become fixed on the surfaces of solids, nourishing grains of quartz for example. It may also undergo nucleation in suspension, leading to a homogenous dispersed colloid (hydrosol), then to a non-rigid gel (hydrogel), and finally to a rigid gel and crystalline forms. Precious opal may be formed by the aforesaid mechanism (Bartoli et al., 1983).

X-ray diffraction analysis shows **opal A** (more or less amorphous), **opal CT** (large peak around 4.1Å indicating the presence of crystobalite and tridymite) and **quartz.** During diagenesis opal is unstable and tends to be transformed into quartz; the presence of opal CT in older rocks (such as pre-Jurassic) must be considered exceptional or at most a late migrant. An optical microscope shows a great variety of facies (Arbey, 1980); besides isotropic *opal, macro-* or *microcrystalline quartz* is also found, as well as the fibrous forms, which are mainly **calcedonite** (negative elongation), **quartzine** (positive elongation) and **lutecite** (positive elongation, oblique extinction). In recent formations fibrous forms are often opal CT but diagenetic recrystallisation does not appear to have affected its microscopic structures, since a few examples have been recognised up to the Precambrian (Siedlecka, 1976). In the geological context, therefore, *it is dangerous to assign a crystalline nature to silicate minerals from microscopic examinations.*

2.5.1.2. Siliceous accumulations and 'recent' silcretes
Siliceous diatoms and sponges live in continental waters, the accumulation of which in a lacustrine environment may give rise to rocks (Meyer, 1984). But this phenomenon is far more limited than the fixation of silica by plants as **phytoliths**. Gramineae (Poaceae), conifers and many other plants contain a considerable amount of silica in their cells. Thus some pedological horizons may include a considerable percentage of opaline phytoliths.

Their forms are varied: needle-shaped, tablets, discs etc. (a complete photographic review is given by Wilding, Smeck and Drees, 1977).

The author has carried out research work in collaboration with specialists on the present-day phytoliths to identify these elements in diverse paleosols; however, no conclusion has yet been reached. Opal phytoliths are relatively soluble and even if not always identifiable in older sequences, they must have constituted a sufficient source capable of sustaining certain accumulations.

In the existing environments silica settles around geyser vents (geyserites, made mainly of opal) or in soil developing on **volcanic ash**. The latter case is less localised than the former: **silan** develops in these soils (Soil Survey Staff, 1975); **andosols** belong to the same category (Duchaufour, 1977, p. 209). On a more modest scale, authigenic quartz crystals have been found in *mangrove muds* (Baltzer, 1970).

Regarding the silcretes: These hard, massive siliceous crusts, which attain a thickness of several metres, presently cover a vast area on the earth, especially in the Sahara, South Africa and Australia. In most cases it is evident that these are fossil silcretes. In Australia for example, two main periods of silicification have been recognised; the ancient one dates to the Eocene and the more recent one to the Pliocene (Langford-Smith, 1978). Except for zones of active volcanism which could account for the origin of a considerable quantity of dissolved silica, it seems that under modern conditions no silcretes are in the process of formation or, at best, the phenomenon is too slow to be clearly perceptible. Contrary to calcretes, geologists do not have at their disposal proper actual models and so in all cases silcretes, which imprint the morphology of many existing deserts, must be viewed *a priori* as fossil accumulations (Goudie, 1985), even if some might be relatively recent (Menillet, 1993).

Smale (1973), based on observations made in South Africa and Australia, distinguished six petrographic types among silcretes, which is evidence of the complexity and great variety of the phenomenon:

— *'Terrazzo' type*: quartz grains cemented by finely crystalline silica; the cement may follow flow structures and may be rich in titanium **(leucoxene)**;

— *'Conglomeratic' type*: pebbles made of 'terrazzo' materials embedded in a siliceous matrix which is often red;

— *'Albertinian' type*: detrital material absent; the rock is similar to the matrix 'terrazzo';

— *'Opaline' type*: detrital material absent in the rock, which is massive, made of opal and chalcedony;

— *'Quartzite' type:* growth of quartz grains imparts to the rock characters of an orthoquartzite.

The classification of facies proposed by Summerfield (1963a, b) is similar to the one discussed above but is more conceptual and perhaps more practical for the investigation of paleosilcretes:

— *Grain-supported* (GS) *Fabric*: detrital quartz grains in contact with each other;

— *Floating* (F-) *Fabric*: detrital quartz grains float in a siliceous matrix;

— *Matrix* (M-) *Fabric*: practically no detrital grains can be identified in the matrix;

— *Conglomeratic* (C-) *Fabric*: a large number of detrital pebbles identifiable in the silcrete.

2.5.2 Alteration of Quartz at Paleosurfaces

The essential feature of the mobilised silica of paleosols and paleosurfaces is the degradation of silico-aluminates. However, quartz itself may be affected in certain surficial environments.

2.5.2.1 Splinter quartz

Microscopic observation of silcretes of the Tertiary of Portugal, in which numerous splinters of quartz have been identified, precludes a regional volcanism for their origin. The same phenomenon is observed in the **violet zones** of the Triassic of Lorraine (Fig. 24). Such splinters probably originate from fragmentation of detrital grains due to *constraints during recrystallisation*.

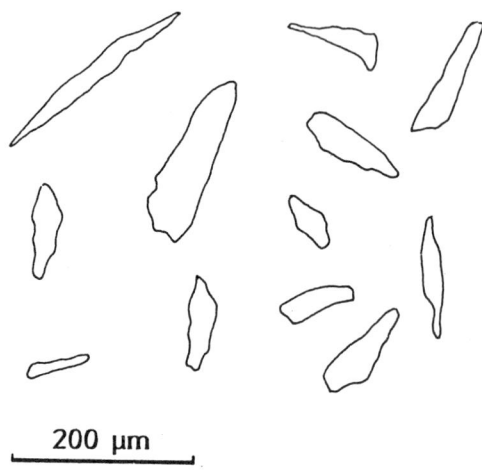

Fig. 24. *Characteristic* **quartz** *splinters in paleosols. They probably originated by fragmentation of the detrital grains in the profile. None of these splinters has the Y-shape characteristic of volcanic splinters. Violet zone of the Triassic in Lorraine.*

2.5.2.2 Corroded quartz grains and desilicification

In carbonated formations diagenetic corrosion of quartz is an established fact; the same may also occur in surficial environments (Fig. 25). A statistical study is necessary however, to prove the mobilisation of silica in alterites; it is practically possible to photograph some grains of corroded quartz in every thin section; a threshold of at least 50% corroded grains is necessary. Where corrosion is normal, the carbonated crusts, particularly in Triassic dolocretes, often enclose quartz grains abraded by rolling, whose surface carry traces of shocks received before their deposition (Durand and Meyer, 1982). This stability may be due to the protection of the surface of the grains by a layer of metallic molecules, such as Al or Fe (Wilding, Smeck and Drees, 1977).

Apart from these localised examples, the phenomenon may assume considerable extension as shown by Thiry, Panziera and Schmitt (1984), who attributed the present relief seen in the *corroded* Fountainebleau Sandstone to the combined effect of growth aureoles and quartz grains. Formation of the **meulieres** (siliceous limestone) of the Paris Basin also presupposes an initial **desilicification** (Menillet, 1984). The formation of itacolumites represents the same phenomenon.

2.5.2.3 Pebbles of weathered quartz

Pebbles of quartzite of the main Vosgian Conglomerate, found in both Quaternary terrains and present-day streams, often lose their original brown colour, become porous, and the haematite which coloured them is rendered soluble, sometimes totally.

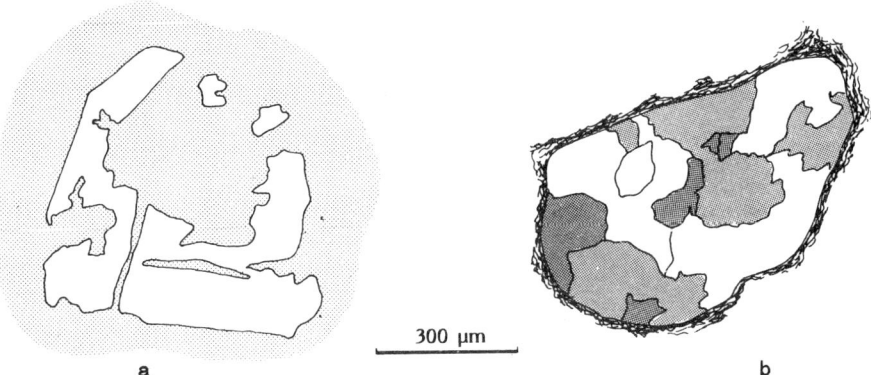

Fig. 25. *Examples of quartz corroded by carbonates*

a) Remnants (in white) of quartz monocrystals embedded in calcite, with a unique optical orientation, in a Miocene channel of the Central Aquitaine. Microscopic view in polarised light.

b) A polycrystalline grain of quartz (in white) in dolomitic crust of the Permian of Vosges. Grain worn due to rolling, embedded in argillo-ferruginous matrix, which has retained its external form but is partially replaced by the dolosparite crystals with a prominent relief. View under natural light.

Some argillaceous layers in the **violet limit zone** of Buntsandstein in Lorraine enclose quartzite pebbles and quartz pebbles weathered at their surface. These alterations are sometimes spectacular, as shown in the profile given in Fig. 26. In a 20-cm thick horizon quartzite and quartz pebbles embedded in an argillaceous (illite) matrix, are *notably dissolved at the upper part while the lower faces are intact* (Fig. 27). The systematic polarity of the phenomenon proves *in-situ* alteration in a **vadose** environment, in which aggressive waters actively circulated, probably downwards.

Towards the base of the pebble layer another facies appears: the brown quartzite becomes *green* at the surface. The change in shade is accompanied by one or many *surficial partings*, tens of millimetres thick, consisting of fine pellicules. The phenomenon initiated 'onion peel' exfoliation of the pebbles, as described by Icole (1974) in the piedmont nappe of the northern Pyrenees. The silica liberated from the pebble layers might be responsible for the origin of silicified lenticles of white cornelian concentrated at the base of the profile.

Nothing justifies the comparison of such a profile with a veritable paleosol but **slickensides** and the polarity of weathering of pebbles implies exundation; the obvious *in-situ* transformation might well be an indication of a fluctuating water table.

2.5.3 Fossil Silcretes: Examples

The profile given in Fig. 27 represents the transition from mobilisation to accumulation of silica, which are the two complementary processes, present at least in a gentle topographic landscape. The mechanisms of siliceous accumulation are of various types.

2.5.3.1 Silcrete in hydrolysing environment

The example cited is taken from the Tertiary in western Portugal, in the coastal basin, south-west of Coimbra. A conglomeratic, sandy-clay formation, which is not accurately dated, is subdivided into three consistent megasequences (Pena dos Reis and Meyer, 1982; Pena dos Reis, 1983). Analysis of sedimentary structures here suggests an alluvial plain

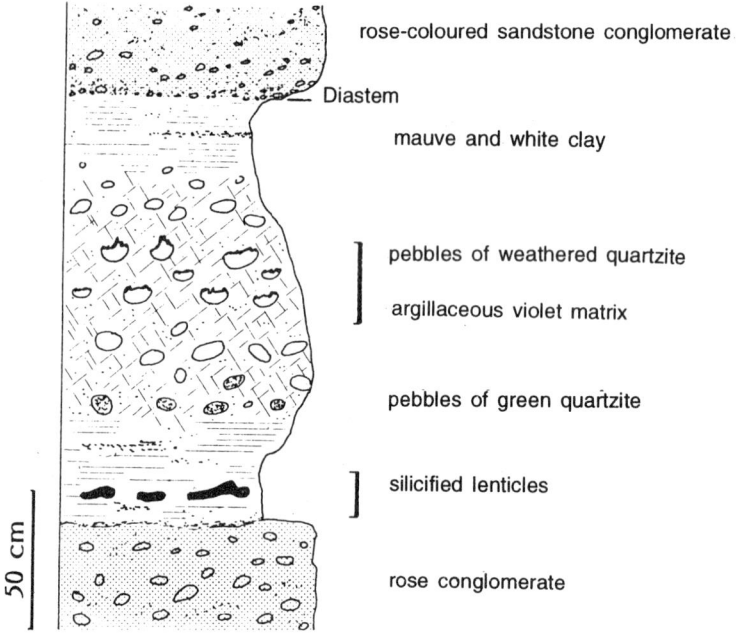

Fig. 26. *Violet limit zone mainly argillaceous in the Buntsandstein des Vosges. The pebbles are weathered on the upper face; green pebbles and silicified lenticles are confined to specific layers.*

traversed by large watercourses, not very deep and pebbly, with a high rate of lateral migration.

The second megasequence, which is more conglomeratic, has a thickness of 30 m. Quartz, quartzite and potash-feldspar together impart to the facies a grey to pink colour, are poorly classified, immature and almost uncemented. This megasequence is terminated at the top by a silicified horizon (1 to 10 m thick) which acts as a good marker horizon for the entire basin. The rock is lithified into sandstone with siliceous cement. **Silicification** is often associated with **ferruginisation** which causes purple or red patches that are broadly columnar in distribution. Well-defined differences distinguish these facies from the underlying sediments:

— *Feldspar is absent.*

— Detrital quartz, often surficially eroded, cemented by a coating of *opal*, comparable to the **silans** of Fig. 28.

— When the facies is ferruginous, **goethite** fills the fissures in the rock and in the quartz grains, giving the grains a split appearance;

— **Kaolinite** is the dominant mineral in the argillaceous assemblages, while illite predominates throughout the megasequence.

Petrographic studies of this silicified horizon (Meyer and Pena dos Reis, 1985) have facilitated recognition of the order of precipitation of authigenic minerals, as illustrated in Fig. 29. **Marmorisation** of sediments and traces of the assimilated root tubules indicate a pedological evolution. Feldspars are destroyed; illite becomes unstable towards the top of the profile and is subsequently replaced by kaolinite. Such an evolution has equivalents

Fig. 27. *Quartzite pebble weathered in the upper part. In the profile given in Fig. 26, such pebbles are located in a specific layer and represent polarised alteration.*

in existing landscapes: the **beginning of lateritic weathering.** This implies a relatively warm and humid climate with redistribution of iron as patches of hydroxides. **Hydrolysis** of silicates is sufficient to make silica soluble, which precipitates locally as opal.

Under these climatic conditions and a gentle landscape the same site may become alternately the seat of mobilisation or **accumulation** of silica, as a function of the drainage potential. **Tectonic stability** of the region during the epoch ensures perenniality of all the mechanisms over a long period of time. This results in widespread crusting with mainly siliceous cementation: silcrete. The second sequence of precipitation of minerals, associated with alunite (Fig. 29), must be considered a later influx derived from the underlying layers which contain small beds of **alunite** and a large number of **boxworks** from which the mineral has probably been removed. These layers are described below.

2.5.3.2 Silcrete in pre-evaporite environment

The sequence in Portugal described in the preceding section is followed by a third megasequence more than 20 m thick and probably of Miocene age. This megasequence is distinct from the preceding ones; it comprises quartz sand within an abundant argillaceous matrix; the colour is locally red, but more often greyish-green.

Three to four silicified layers have been identified in this unit, each less than 2 m thick; their extension, however, is limited to correlation at the scale of the basin. These silicified horizons have been locally transformed into quartzites; they are always bleached and bioturbated, the **burrows** constituting the channels through which the silica diffused into the sediments. Under an optical microscope flaky *fibrous silica* (**quartzine**) occurs

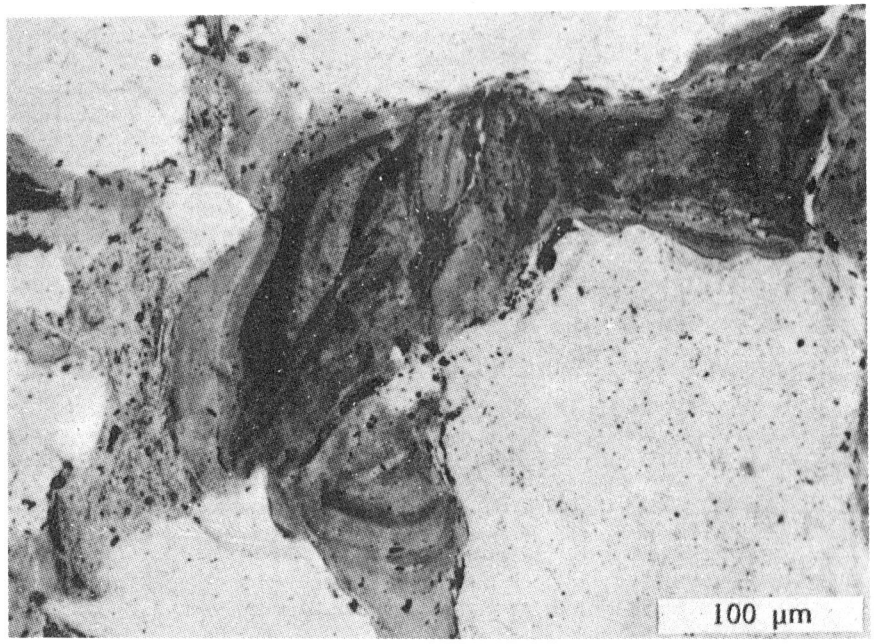

Fig. 28. *Silan in the Eocene silcrete of Perigord. Quartz grains are cemented in the deposits of finely laminated silica. The grains are diverse in colour due to oxides. Microscopic view under natural light (photograph, F. Daugas).*

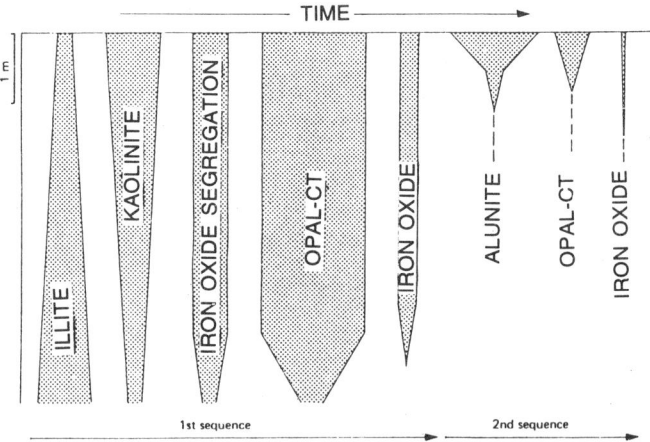

Fig. 29. *Order of precipitation of authigenic minerals in a* **silcrete**. *The horizontal line at the top of the diagram represents the top of the silcrete. The second sequence of precipitation corresponds to an influx of minerals from the underlying layers. Tertiary, Portugal (Meyer and Pena dos Reis, 1985; with permission from the Journal of Sedimentary Petrology).*

between the grains of quartz, as a result of the epigenesis of argillaceous minerals; the microstructure is of the **skelsepic** type but rhombohedral boxworks are also very frequent. A schematic evolution with respect to the granulometry and degree of silicification of the argillaceous assemblages is shown in Fig. 30.

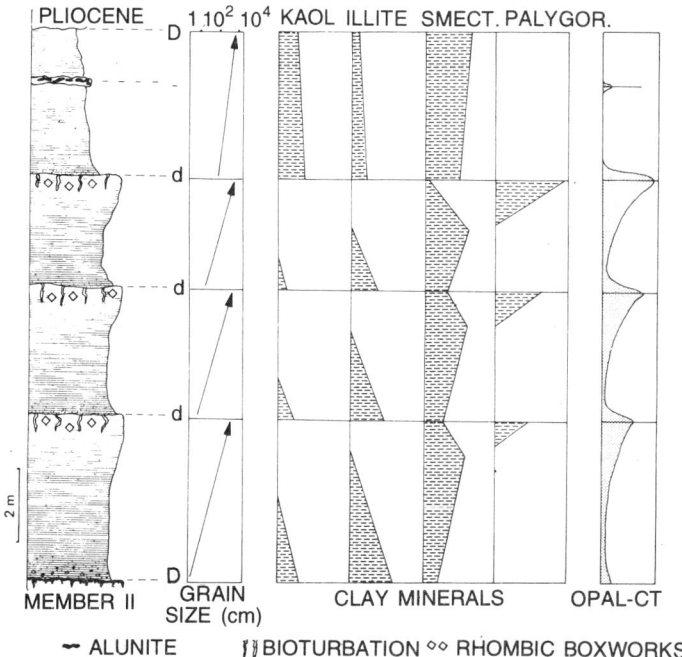

Fig. 30. *Silicification in pre-evaporite environment, Tertiary, Portugal. The top of each detrital sequence is intruded by bioturbation, palygorskite and opal (Meyer and Pena dos Reis, 1985; with permission from the Journal of Sedimentary Petrology).*

The duration of each **diastem** is long enough to allow considerable chemical transformation. Illite and kaolinite disappear, while neoformation of **palygorskite** takes place. This evolution may be associated with basic lakes in a relatively dry climate (Millot, 1964, p. 200), but the general prismatic structure and numerous **slickensides** prove the emergence of sediments, at least temporarily; these structures are typical of soils rich in swelling clays. The silica responsible for the fossilisation of these structures subsequently invaded this unconsolidated alterite. The paleoenvironment was an alluvial plain in which watercourses flowed almost at the base level (Pena dos Reis, 1983). *Chemical conditions* were probably highly *unstable* depending on the water available (Meyer, 1981): increase in pH values (dryness) caused destruction of some particular silicates and neoformation of palygorskite, while lowering of pH (flood) precipitated dissolved silica. Interpretation of the boxworks, which are very small, is critical; there is an absence of carbonate in the formation and the boxworks could be the vestiges of alunite, or even of pyrite, the minerals which could have liberated sulfur, which may have migrated towards the lower levels of silcretes described in the preceding section.

This continental silicified layer also merits the name silcrete but it may be noted that the paleoenvironment was very different from the one which gave rise to the silcretes described under Sec. 2.5.3.1.

2.5.3.3 Paleosequence with silcretes

Studies conducted on the continental Tertiary facies in northern Perigord have facilitated understanding the later changes in the facies (Daugas, 1981; Daugas and Meyer, 1982). The detrital complex of Saint-Pardoux, of Middle to Upper Eocene, may be virtually reconstructed from the border of the basement near Saint-Pardoux-La-Riviere (**upstream zone**) to the west of Brantome, about 20 km south (**downstream zone).**

This detrital complex is characterised by relatively fine facies at the base, becoming coarser at the top, indicating two different episodes of sedimentation. Towards the upstream, the coarse erosional products of ferruginous crusts (**siderolitique**) and of the basement complex are encountered; downstream, the formation is finer, more mature and thicker (Fig. 31). Significant silicification, though discontinuous, affects both the facies, which, however, differ in chemical characteristics, as evident from their argillaceous assemblages: only kaolinite has been detected in the upstream domain, while in the downstream domain kaolinite is accompanied by smectite and even by palygorskite (as well as a few crystals of gypsum). The neoformation of these minerals in the downstream domain may be viewed as an indicator of silicification.

Silcretes can thus form in an *acid environment* if the drainage is good, as well as in a *basic environment* when waterlogging has occurred.

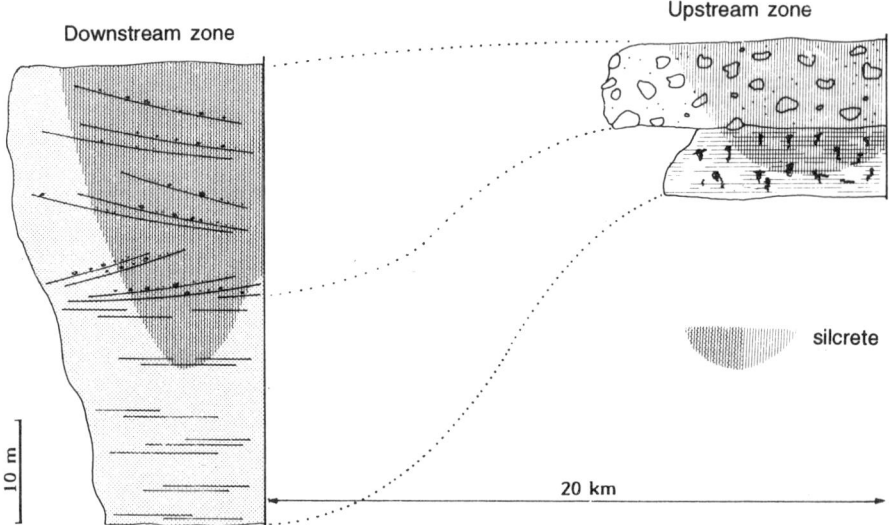

Fig. 31. *Tertiary paleotoposequence, silcrete, north of Perigord. Upstream, where silicification affected argillaceous facies and conglomerates, kaolinite is the only clay mineral; in the downstream facies of fine sandstone, kaolinite is accompanied by smectite, even palygorskite. Silicification is mainly of opal-CT (quartzine under optical microscope).*

2.5.3.4 Influence of burial diagenesis on silcretes

The fossil silcretes described above are all relatively recent. Silcretes have also been identified in ancient sequences, for example the **violet limit zone** (VLZ) of the Lower Triassic of Lorraine (Durand and Meyer, 1982).

The VLZ attains a thickness of a few metres (Fig. 32a) and is a marker bed, almost continuous at the top of the main Conglomerate, which is an oligomictic pudding of pebbles of quartz and quartzites. Overlying the VLZ, sandstone belonging to Intermediate Beds are feldspathic and rich in phyllites; they enclose siliceous elements derived from the VLZ, which proves the anteriority of silicification. It grades laterally into **dolocretes** or into layers rich in variegated clays comparable to those schematised in Fig. 26. In its most developed form silcrete is a massive bed of red, grey or white carnelian, in which a few elements of the parent rock—pebbles and grains of quartz—exist. The VLZ has long been interpreted as a paleosol; in fact, the Germanic sea lay north-eastwards in that epoch. The depth of burial of this ensemble is not negligible, being at least 1500 m. The *paleoenvironment* for the formation of these facies may be reconstructed only indirectly, either by studying the **structures, relict minerals** and **boxworks** fossilised by silica, or by investigating the *lateral gradation* towards the facies in which siliceous impregnation is discontinuous.

Such an analysis led to detection of the precursors of silica detailed by Durand and Meyer (1982). Only the most important features are described here.

— Numerous silicified rhombohedral structures, similar to **ribboned pellicules,** suggest the possibility that silica had invaded a carbonated crust in many cases. The presence of very small relics of **dolomite,** which could be dolocrete, has been determined by microprobe analyses.

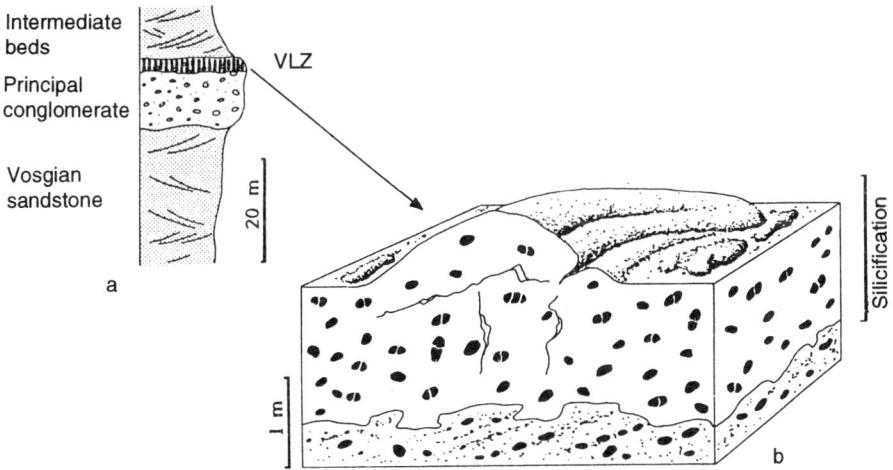

Fig. 32. *Silicified* **violet limit** *zone (VLZ) in Bundtsandstein des Vosges.*

  a) Stratigraphic position of VLZ.

  b) Upper limit of silcrete with swellings. Quartz pebbles (in black) are often fractured by the silicified mass, even diluted in the siliceous matrix perpendicular to the swellings.

— Pseudomorphs of gypsum and anhydrite with chickenwire structures indicate the presence of a **pre-evaporite** regional environment (see Fig. 21).

— Optical microscope studies enable delineation of the progressive gradation of an argillaceous matrix into a matrix of fibrous and flaky silica (Fig. 33).

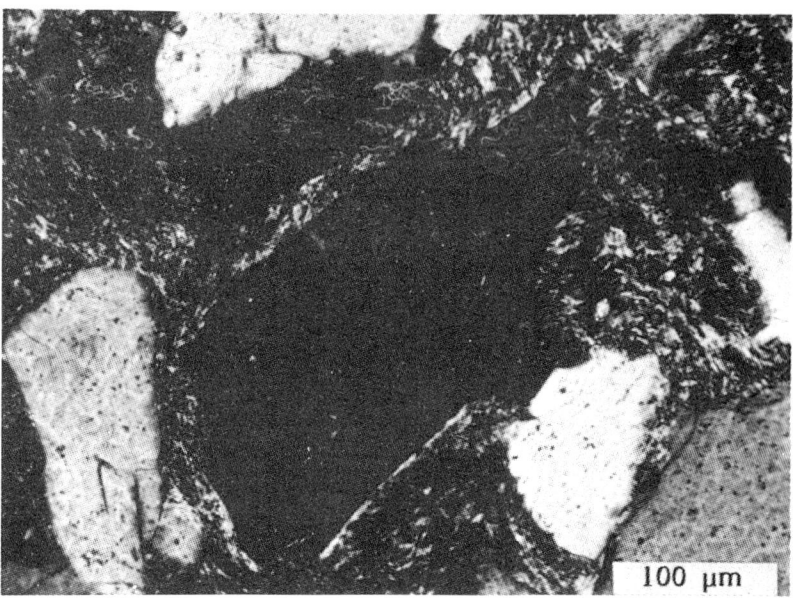

Fig. 33. *Fibrous and flaky siliceous matrix. These facies are common in silicified violet zones, which appears to be the result of silicification of an argillaceous matrix. Buntsandstein des Vosges; microscopic view under polarised light.*

— Quartzite pebbles often have large fractures filled with a silicified matrix (Fig. 34). Fissuring and opening up of cracks are the result of growth of crystals of precursor minerals such as gypsum or calcite.

— The top of the silcrete is often swollen (see Fig. 32b), causing vast supple deformations similar to the mechanism observed in present-day **vertisols,** wherein alternation of desiccation and humidification mix the argillaceous materials, with the occasional appearance of **gilgai** microrelief.

The essential elements of interpretation may be formulated on the basis of the structures of minerals which preceded silicification. A great variability is observed in the facies, this differentiation being favoured by a particular paleogeography—a *very flat alluvial plain* away from the source region. A small variation in the topography induces the appearance of *very diverse microenvironments.* Thus a landscape can be visualised in which sediments evolved in pedoclimates, linking the *extremely dry* (carbonated crusts, **solonetz**) to the highly humid (*vertisols*, or even ponds).

Burial diagenesis has left imprints in all the facies, as evidenced by the mineralogical composition:

— X-ray diffraction analysis shows that silicification is made up exclusively of **quartz** but under an optical microscope quartzine is also observed.

EXAMPLES OF PALEOALTERITES AND PALEOSOLS

Fig. 34. *Quartzite pebble cut into sections in a silcrete. The fissures in the pebbles are filled with silicified sandstone matrix. Buntsandstein des Vosges; sample from M. Durand.*

— **Illite** is the only clay mineral present, in contrast to the variety of facies and the argillaceous paragenesis found in present-day landscapes.

— Oxidised iron enters only one mineral, namely **haematite**, which is very different from the complex melange of common hydroxides in present-day continental environments.

Hence a paleogeographic reconstruction cannot be achieved directly on the basis of these facts; it is necessary to integrate the transformation related to diagenesis.

2.5.4 Fundamental Types of Paleosilcretes

It is difficult to give a precise **paleogeographic reconstruction** based on purely descriptive types (Sec. 2.5.1) identified in older sequences. This has led many authors to interpret silcretes as a function of the associated facies or facies invaded by silica (Thiry and Milnes, 1991).

2.5.4.1 Silicified conglomerates and sandstones

Silicification of conglomerates may differ markedly from those described earlier from the Triassic, for example the conglomerates of Nemours (Thiry, 1981, p. 122). This reworked Sparnacian formation of flint pebbles originating from chalk is terminated at the top by a silcrete (Fig. 35). The profile, from bottom to top, is as follows:

— Development of finely stratified *sand cappings* (Fig. 36) in the upper part of each pebble (nearly 2 m thick);

— A *lustrous pudding*; in addition to the silicified cappings, the matrix of the pudding becomes lustrous grey ochre (2.5 m).

— A *quartzite pudding*; the matrix recrystallises into visible quartz; flint and its cappings are partly replaced by quartzitic cement (1.5 m).

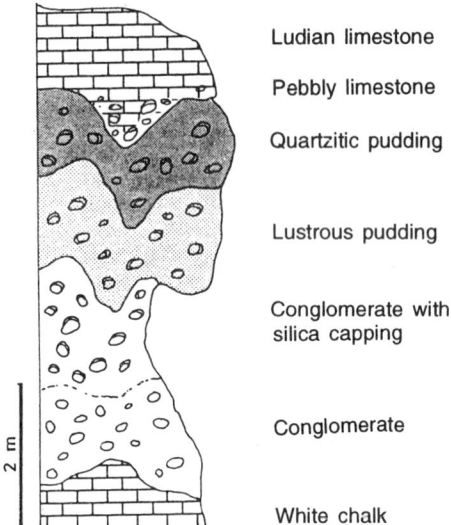

Fig. 35. *General disposition of the facies of silicification in the conglomerates of Nemours (after Thiry, 1981).*

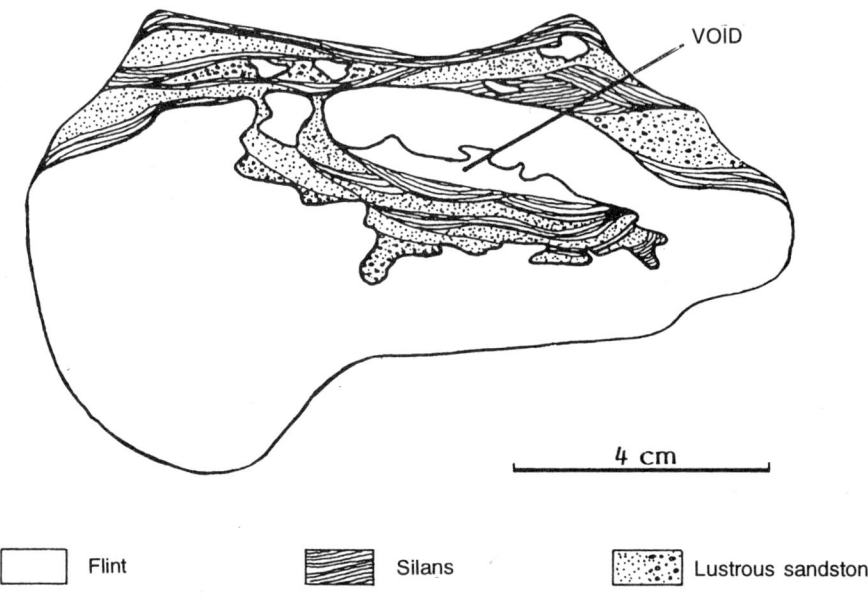

Fig. 36. *Flints covered with largely silicified cappings. The large internal cavities of alteration contain the same deposits as the cappings. Silans are quite rich in oxides of titanium, found as granules in glossic sandstone. Silcrete at the top of the conglomerate of Nemours, Sparnacian (after Thiry, 1981).*

— Eventual *calcification* of the surface of silcrete at the time of deposition of the overlying Ludian limestone.

The speciality of such silcretes is the silicified cappings of pebbles, emplacement of which depends on the mode of percolation of solutions.

Transformation of sandstone into **quartzite strata** has been thoroughly studied in the Albian of Gard (Parron et al., 1976; Guendon and Parron, 1983). The formation, a glauconite-bearing sandstone, underwent weathering initially. Glauconite disappeared with the appearance of **kaolinite** and **goethite**, the rock lost its coherence and became red-coloured sand. This red sand was further transformed towards the top into a very hard quartzite stratum, whitish-pink in colour and 10–12 m thick. The aforesaid authors have interpreted the phenomenon on the basis of rain-water: silica, compared to all other cations, readily dissolves in rain-water and attains the limit for *precipitation of quartz*, while other minerals continue to be dissolved at a slower pace; kaolinite disappears, goethite is also dissolved and *iron removed from the profile*. Such an evolution requires a leaching environment, unsaturated and relatively acidic, i.e., not unlike formation of the silcrete associated with lateritic or siderolitic facies (Langford-Smith and Watts, 1978; Twidale, 1983; Thiry et al., 1983).

2.5.4.2 Silicified clay

Clay is almost always present in the deposit undergoing silicification but it is generally the argillite layer which undergoes total **epigenesis** by silica minerals. This is the case at the base of the profile just described in the Tertiary of Perigord (see Fig. 31).

Thiry (1981, p. 104) has discussed the formation of a silicified profile at the top of Sparnacian plastic clay; he recognises from bottom to top:

— argillite horizon (mainly kaolinite), with siliceous granules (65 cm);
— silicified horizon, columnar, prismatic, more or less dislocated (75 cm);
— silicified pseudonodular horizon (35 cm);
— pseudobrecciated horizon (40 cm); argillaceous assemblage dominated by illite.

Menillet (1984) considered a similar 'clay with siliceous limestone' as the material likely to give rise to many siliceous limestones of the Paris Basin.

2.5.4.3 Silicified calcretes and dolocretes

More or less discrete silicification of recent calcretes has been described (Conrad, 1969; Kulke, 1974; Netterberg, 1974; Arakel and McConchie, 1982). A comparable example from the Thanetian of Catalonia has been shown in Fig. 37; Valleron (1981) described a similar example from the Eocene of south-east France.

It is also to be noted that carbonated crusts act as a **precursor** to the accumulation of silica, for example from the Triassic (Röper and Rothe, 1975; Davaine, 1980) and from the Permian (Meyer, 1981), or even from the Precambrian (James et al., 1968).

This phenomenon is not exceptional since it is a known fact that a simple lowering of pH value (dilution, decomposition of organic matter) destabilises the carbonate and precipitates the silica available in the solution.

2.5.4.4 Silicification in pre-evaporite environments

It is generally observed that silicified layers are associated with formations having pre-evaporite indexes. However, in most cases they are ascribed to a **lacustrine** or **palustrine** environment, usually with a specific chemistry (Surdam, Eugster and Mariner, 1972; Milliken, 1979; Crouzel and Meyer, 1983). The second type of silcrete described from the

Fig. 37. *Partial silicification of a fossil calcrete. The silicified laminae are clear and shown by arrows. The pencil indicates the scale. Upper Thanetian Catalonia (photograph of outcrop presented by P. Anadon).*

Tertiary of Portugal is definitely similar to these. In the **reconstruction of paleogeography** they delineate a **paleosurface**.

Silicification associated with an **evaporite** environment indicating emergence is more pronounced. Silicified gypsum rosettes in the Ludian of Champagne (Sec. 2.4.2) are of this type. Arakel and McConchie (1982) have found comparable indexes in the paleosurface outcrops of many places in parts of Australia. Rubin and Friedman (1981) described similar phenomena from the Cambrian-Ordovician. Silcretes from the Triassic of Lorraine probably indicate *complex hydraulic paleoregimes* in which the pre-evaporite trend was not constant, disappearing temporarily and locally.

2.5.5 Problems of Diagnosis of Fossil Silcretes

Fossil silcretes in ancient sequences are not really rare; though silica appears to be only slightly soluble in the prevalent environment, the duration of geological processes is sufficient to mobilise and fix it at certain favourable sites in the landscape.

It is usually possible to distinguish between paleosilcretes and silicified layers of diagenetic origin in marine sequences (flint or chert layers) by considering the general historical background and paleontological criteria. The distinction is more difficult between lacustrine silicification and that related to paleoalterites. The best criteria in this case are, perhaps, the complexity and variability of the facies: *The more varied and complex they are, the more significant the role of weathering.*

Paleogeographic reconstructions on the basis of silcretes require identification of the facies which have undergone silicification. When there is a simple **cementation** by silica, the approach is easy; it becomes more delicate as evolution proceeds with **epigenesis** even if silica in this case maintains the forms and structures of the minerals very minutely. The eliminated evaporite minerals may be identified on the basis of the fibrous form of silica as **quartzine** and lutecite (Cayeux, 1929; Folk and Pittman, 1971). This technic should be used with prudence. Quartzine appears effectively as a replacement product but together with evaporite minerals it may replace carbonates and argillaceous minerals (Durand and Meyer, 1982). *These fibrous minerals therefore provide information only about the chemistry of solutions*, probably very basic, from which the silica precipitated (Arbey, 1980).

Some silcretes are rich in titanium (**leucoxene**). This is particularly prominent at the base of the profile in juvenile horizons and towards the summit of degraded facies (Laville, pers. comm.). This phenomenon is equally well known in the paleoalterites of Brittany and in the Tertiary of the Paris Basin (Thiry, 1981), as well as from Australia (Langford-Smith, 1978). However, the phenomenon is not common, only occasional, because it is a matter of simple relative concentration when the parent rock itself is rich in titanium.

SILCRETES

Morphology and Facies
Discontinuous and irregular crust, grey to white in colour, or varying with the iron oxide content; normally maintains the facies preceding silicification.

Microfacies
Quartz, microquartz, fibrous forms (chalcedonite, quartzine etc.) and isotropic opal in recent formations.

Mineralogy
Quartz, opal A and opal-CT which recrystallise as quartz during diagenesis; eventual abundance of titanium oxide (leucoxene) in juvenile horizons and horizons of degradation; iron oxide tends to be eliminated by silica.

Substratum
Sandstone, conglomerate, argillite, calcrete etc.

Frequency
Numerous paleosurfaces dating from the Precambrian to Recent.

Climate
Warm and dry evaporite environment (basic process) or more humid and hydrolysing environment (acidic process).

Tectonic context
Slow rate of formation of silcretes in hydrolysing environment implies tectonic stability.

Convergence
Marine silicification but more regular than silcretes.

2.5.6 Principles of Interpretation of Fossil Silcretes

Paleosilcretes constitute a complex assemblage, their facies often showing contradictory influences (Summerfield, 1983c). The paleoenvironment could have changed at the scale of landscape, that is, at the scale of a few tens or hundreds of metres; it might have also changed in time at a seasonal scale even more effectively since *different pedoclimates* succeed each other over the long period necessary for silicification. Most of the known cases appear to correspond to a series of paleoclimates evolving between two opposite poles however, acidic and basic.

Essentially acidic process: An environment with **hydrolysing** climate slowly and regularly mobilises silica, which proceeds towards the sea if drainage is good. If drainage is relatively restricted, silica is fixed at the base of the profile or at favourable sites. The process is slow, which favours the growth of quartz and cementation. The rate of sedimentation must be *slow* and **tectonically** relatively stable.

Essentially basic process: The environment shows an evaporite trend, with a relatively higher pH value, which may *mobilise silica in large quantities*. It precipitates locally by a simple lowering of pH value in the environment (dilution due to rain-water or inundation). This process is more rapid than the previous one. Silica crystallises mainly in the *fibrous form* (quartzine and lutecite) but quartz may also appear, which is rich in **carbonate** or **sulfate** inclusions. The climate is *deficient in water,* even though the water table may rise to the surface locally and temporarily.

2.6 Oxidised Paleoalterites

Chemical elements are fixed in soils preferably as oxides and hydroxides: the major elements are iron and aluminium and the minor elements are manganese, titanium etc. An association of Fe and Al is common in **lateritic soils** and in **cuirasses** but is not always systematic. For the sake of clarity, it is necessary to consider paleoalterites rich in Fe and Al separately.

2.6.1 Ferruginous Paleoalterites

2.6.1.1 Iron in present-day alterites

In continental environments iron may eventually enter into carbonates (Sec. 2.3), sulfides or sulfates (Sec. 2.4). Iron liberated during alteration of primary silico-aluminates is less mobile, however, and is fixed in the profile in organo-mineral complexes in the form of amorphous minerals, then as oxyhydroxides and oxides. The most common minerals are given below (Schwertmann and Taylor, 1977; Guillet and Souchier, 1979):

Limonite is a melange of small crystallites composed of the following minerals:

Goethite: α-FeO(OH) is an oxyhydroxide commonly present in the soil of all climates. It gives a brown colour to the horizon and becomes very dark when present in abundance.

Haematite: α-Fe$_2$O$_3$ eventually accompanies goethite in soils developed in **Mediterranean, tropical** or **equatorial climates.** It produces an intense pigmentation and colours the horizon red when the size of the crystallites reaches the scale of micrometres; the **colour** becomes mauve with recrystallisation due to the growth of crystallites (Durand, 1975). With en masse crystallisation, the colour becomes very dark.

Lepidocrocite: γ-FeO(OH), of orange colour, characterises hydromorphous soils. Like *ferrihydrite,* it is of little geological interest. These minerals are unstable and tend to recrystallise in the soil *a fortiori* during diagenesis (Fig. 38).

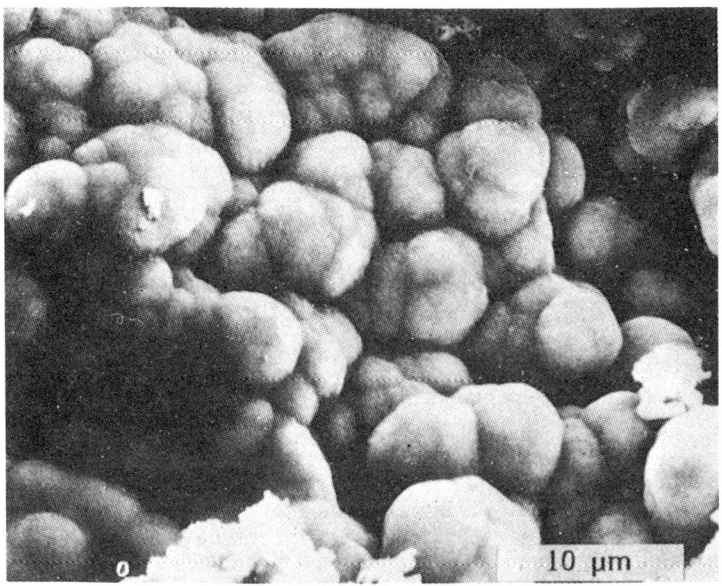

Fig. 38. *Lining of goethite in the pores of Wealden paleosol. The globular appearance suggests that the mineral resulted from recrystallisation of an amorphous deposit. Observation under SEM.*

The mode of redistribution of iron in the soil varies according to the parent rock, climate and drainage conditions. Three processes appear to be essential.

Rubefaction: The parent rock undergoes alteration with complete dissolution of carbonates. Liberated iron is fixed on the surface of clay minerals in the form of small oxide crystals, which impart an almost uniform brownish-red colour to it. Terra rossa is a good example. The phenomenon involves a very permeable alterite, relatively heavy rainfall and periods of drought, making the conditions favourable for the dehydration of iron oxide (Duchaufour, 1977, p. 409). The alteration, producing an ochreous tinge, may affect a sediment which is not part of a pedological profile. For example, in the Sinai desert dune-sand becomes progressively darker in colour from the most recent dunes on the Mediterranean coast to the older ones in the interior of the continent. This evolution, visible to the naked eye, has been confirmed by analytical studies (Williams and Yaalon, 1977).

Marmorisation and the formation of **plinthites.** Segregation of iron is found in sediments subjected to temporary hydromorphy, which is concentrated in definite well-aerated zones. *Marmorisation* commences with the appearance of more or less vertical rust patches that are often in the shape of 'tongues' or **glosses.** This phenomenon is common in diverse temperate climates.

A comparable evolution, but more pronounced, may take place in warm and relatively humid climates. A marmorised red-mottled horizon forms at the limit of water-table fluctuation, i.e., a *plinthite*. The content of iron is due to *in-situ* redistribution and also later supply (Flach et al., 1969). If the water table is depressed, an irreversible hardening of the horizon leads to the formation of a *petroplinthite* in which goethite gives way to haematite.

Cuirasse formation. Like silicification, formation of an iron cuirasse requires considerable time to attain a substantial thickness. The term 'contemporary' is not strictly applicable to **cuirasses** although a few may be recent. The evolution observed in the kaolinic sandstone of Senegal sums up the essential features of the phenomenon (Nahon, 1976; Nahon and Millot, 1977). These authors distinguish the following stages:

— *Simple sandy facies* (Fig. 39a): parent rock composed of 90% quartz, 8% kaolinite and 2% goethite, retains its texture but evolves into a sandstone with 60% quartz and 40% oxyhydroxides of iron.

— *Mixed facies* (Fig. 39b): differentiation into coarse nodules, producing a rock composed of 40% quartz and 60% oxides of iron (aluminous haematite with 13% substitution by Al_2O_3 molecules).

— *Pseudopisolitic facies* (Fig. 39c): division of nodules and recrystallisation of aluminous haematite produces very dense pseudopisolites: these are made up of aluminous goethite, 16–22% substitution by AlO(OH) molecules.

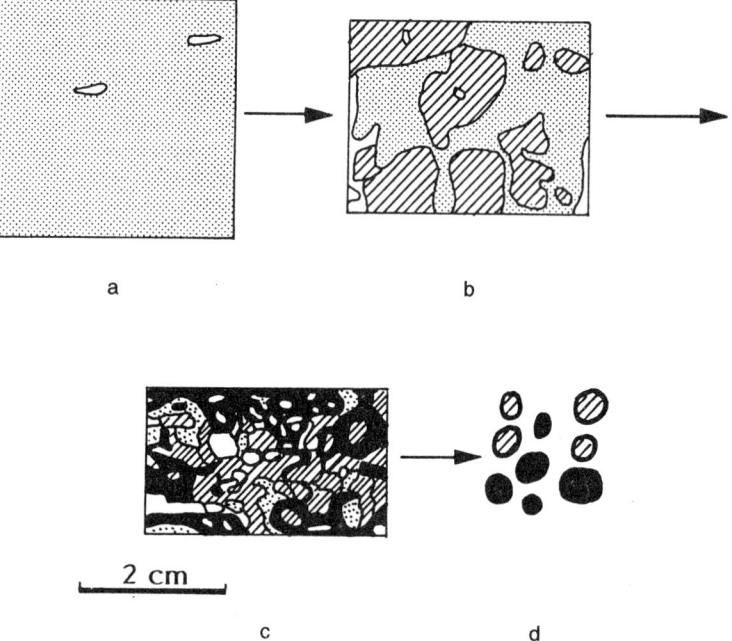

Fig. 39. *Stages of evolution of a ferruginous* **cuirasse** *on kaolinic sandstone (after Nahon and Millot, 1977).*

a) Simple sandy facies;
b) Mixed facies (differentiation into nodules);
c) Pseudopisolitic facies;
d) Free pseudopisolites (erosion of cuirasse).

— *Free pseudopisolites* (Fig. 39d): ferruginous nature of the cuirasse is progressively cut off by fissures and tubules isolating the pseudopisolites.

Geomorphologically, these cuirasses tend to sink in, leading to the epigenesis of unweathered rocks or underlying alterites, and themselves degrading into fine gravel facies unconsolidated at the surface (Fig. 39d).

2.6.1.2 Rubefaction and formation of red beds

The red beds described in geological literature represent varied sedimentary environments: aeolian, fluvial or even deltaic (Turner, 1980). However, *red beds are poor indicators of climates*; although a desert or subdesert paleoenvironment is more common, red beds associated with a sufficiently humid tropical environment are also known to occur. Although a continental environment is almost always involved in the origin of the red coloration, there are strong reasons for arguing that no unique process is responsible for the formation of the red series. Buntsandstein in north-eastern France is a good example: it is red from the Vosgian sandstone to the Votzia sandstone in spite of varied sedimentary environments (Durand, 1978) and climatic fluctuations which have been proved by paleontological investigations (Gall, 1971).

Two fundamental propositions may be noted for the red beds:

— For the iron contained in the sediment to be oxidised and to remain in an oxidised state during burial, the *sediment must not contain organic matter;* this may be due either to its absence in the depositional environment or because it itself rapidly **oxidises.**

— The red colour observed in ancient sequences related to the presence of haematite is brighter, mainly due to the presence of goethite or its melanges, than that observed in recent sequences; thus the red coloration of the sediment is *simply induced in the depositional environment* and assumes its definite character on **diagenesis during burial** (Walker, Wagh and Crone, 1978; Hubert and Reed, 1978).

2.6.1.3 Traces of hydromorphy fossilised in ancient sequences

This phenomenon is observed in formations of various ages. For example, the Oligo-Miocene molasse of Aquitaine, the Permian sandstone of Vosges and the Siluro-Devonian Old Red Sandstone of Wales attest that traces of hydromorphy can be fossilised by redistribution of iron. A variety of colours—red, mauve, rusty-brown and grey—can be differentiated in the outcrops and some typical features consistently found.

— Nuances of coloration are perfectly horizontal and independent of the existing topography. They extend laterally for tens of metres. The **effects of fluctuation** of the water table on the depositional environment are clearly visible. These features are good criteria for recognising the existence of a **vadose** zone, consequent to **emergence.**

— Clear sinusoidal strips vertically intersect the sediment; they affect a single horizon and can be assimilated to *glosses*. They represent a *pedogenesis*, viz., decoloration develops following evaporation along fissures and traces of roots.

— Stains of various colours are juxtaposed with no particular arrangement: the horizon is *marmorised*. These less characteristic facies are intermediate between the preceding types; they may represent paleosols with temporary hydromorphy (Buurman. 1980; Retallack, 1983). Such facies are common in the ancient sequences and geologists have proved the existence of hydromorphous paleosols in them. But this assimilation is suspect; *these traces may develop independent of any pedogenesis*. In a sequence observed in outcrops **marmorisation** has had an equal opportunity to have developed in the outcrop environment as well as in the depositional environment. To conclude, there are many cases in which marmorisation cannot be interpreted.

2.6.1.4 Paleosols with plinthites

A fossil paleosol in the Valanginian of the Wealden facies in Meuse has already been discussed as a site of formation of kaolinite (Sec. 2.2.3). Towards the top of the same areno-argillaceous formation, other paleosols present a different aspect. The base of the bioturbated horizon with root traces is marked by *irregular brown hard layers a few centimetres thick*. These are accumulations of goethite and the rock locally contains more than 45% Fe_2O_3; this mineral is concentrated along the pedotubules, mainly in strings of horizontally flattened irregular nodules (Meyer, 1976). The **nodules** generally show concentric aureoles marked by slight changes in colour. A microscopic study shows goethite preferably penetrating in argillaceous **plasma** and causing a partial epigenesis of quartz grains.

The possibility that these facies originated due to the oxidation of sulfide layers, either before burial or after outcropping, cannot be excluded *a priori*. The first hypothesis may not be very probable. In fact, the Barremian transgression, which conceals the Wealden, implies a progressive rapprochement of the coastline; it caused the evolution of an environment which was initially oxidising then became more and more reducing. The second possibility is still less plausible because the base line of the marine transgression reworks the fragments of concretions and the goethetic plaquettes.

Van Wambeke (1973) proposed a model to explain such horizons in present-day **alluvial soils** in the grassy plains of Colombia. The topography is very flat with poor **drainage**; there is little migration of iron, which is fixed in the zone of water-table fluctuation. A horizon rich in goethite (plinthite) develops, which may harden during the process of desiccation.

2.6.1.5 Ferruginous paleocuirasses

2.6.1.5.1 *Fossil cuirasses in present-day landscape.* The series of facies in paleocuirasses and the mechanism that might have been responsible for their formation have been particularly well studied on **cuirasses** that are no doubt fossils but always occur at outcrops. This is notably the case in West Africa and based on the work of Leprun (1979) the following model can be proposed.

The formation of this cuirasse, albeit not accurately dated, is certainly pre-Quaternary. This has been equally well recognised in Senegal, northern Cameroon and Hoggar. A comparative study of many outcrops resulted in the construction of a profile synthesised from them (Fig. 40) consisting of many layers, described below from top to bottom.

Cuirasse hard, nodular, with variegated matrix and predominantly red. It underwent a *relative accumulation* of iron, concurrent with the withdrawal of fine particles towards the base and the elimination of chemical elements in solution (all the alkali elements, part of the silica and also some aluminium). This relative accumulation was effected by the precipitation of goethite. **Kaolinite** was destabilised in order to liberate aluminium, which did not form **gibbsite** but entered as a *substitution in goethite* (up to 20%). At the top of the cuirasse all the minerals underwent a phase of similar dissolution; liberated iron reprecipitated as non-aluminous haematite and there was *absolute accumulation*. The

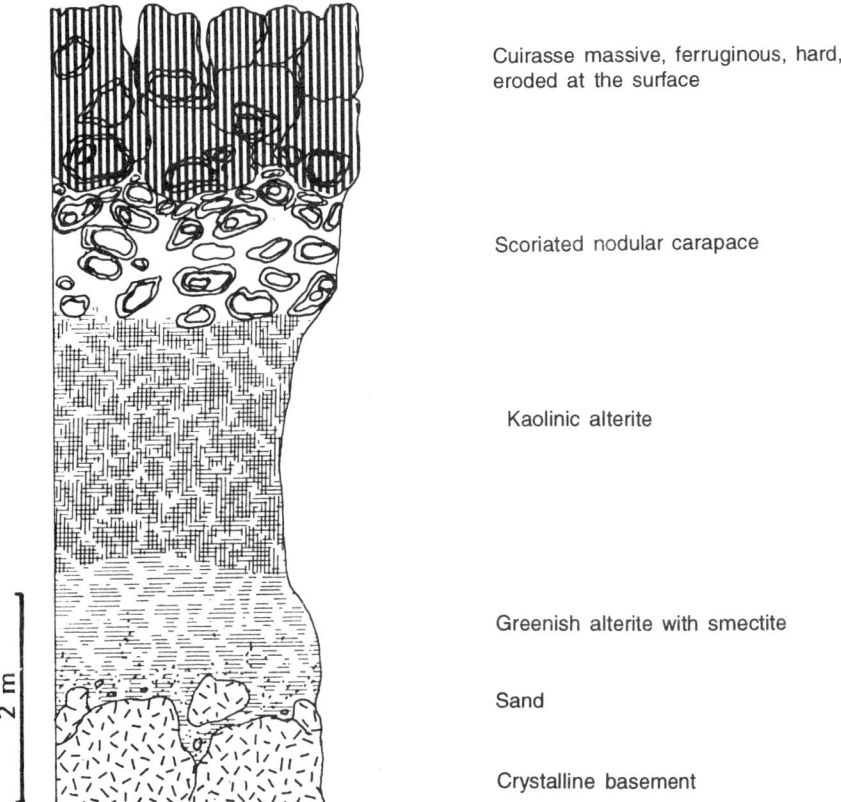

Fig. 40. *Paleoalterite with ferruginous cuirasse (after Leprun, 1979). This profile was constructed by synthetising many known sections on crystalline rocks in West Africa. The thickness of the profile varies considerably.*

present cuirasse is no longer in equilibrium with the climate, tends to degrade at the surface and is broken down due to erosion into blocks.

Carapace with ferruginous nodules enveloped in a red argillaceous and fairly indurated matrix; it may be considered a *transitional stage* wherein the facies are similar to the cuirasse but less evolved.

Kaolinic alterite highly argillaceous, variegated or spotted. The characteristic paragenesis is kaolinite, goethite (practically without substitution) and quartz. This layer may undergo silicification. Silicon, liberated by weathering of silicates in the cuirasse, may precipitate as **opal** or **quartz** when the Al in the profile is not sufficient for neoformation of kaolinite.

Greenish 'pistachio' alterite rich in unweathered or barely weathered primary minerals and authigenic smectite. The geochemical balance indicates that there was a moderate loss of elements including iron from the assemblage. This alterite is discontinuous and preferentially located in depressions with poor drainage. The weathering which gives rise to smectite follows kaolinisation in time; this corresponds to transformations evidenced in vertisols presently developing in the same region.

A chronology of the appearance of diverse facies in these paleoalterites can be established. The cuirasse, carapace and kaolinic alterite are evidence of the **ferrallitisation** characteristic of **warm and humid climates**, which no longer prevail, and therefore they are fossils. The **lithorelics** present in the carapace and cuirasse but absent in the subjacent kaolinic alterite, again indicate that the cuirasse developed directly from the subjacent **parent rock** and therefore prior to the kaolinic covering on the former. Greenish alterite is a late product, almost in equilibrium with the contemporary dry climate (of Sudan) prevalent in the region; this same climate tends to degrade the cuirasse at the surface. The chronology of facies appearance is schematised in Fig. 41.

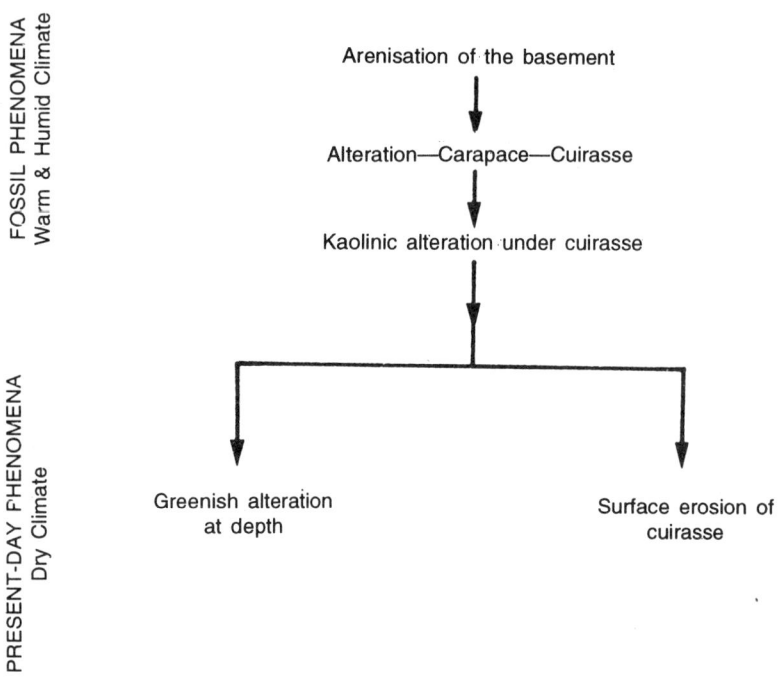

Fig. 41. *Chronology of events which led to the formation of a ferruginous cuirasse in West Africa (after Leprun, 1979).*

The same study showed that the phenomenon of formation of a cuirasse varies in *intensity, facies* and *vulnerability to degradation*, according to the subjacent **parent rock**. The importance of this fact in prospecting should not be overlooked by the reader (Fig. 42). The common presence of lithorelics in cuirasses also shows the **autochthony** of the **weathered materials**. A study of several **toposequences** revealed the great diversity of facies and profiles. The present-day outcrops are schematised in Fig. 43. These *quick changes* are likely to be encountered in the facies as such paleoalterites were fossilised under a sedimentary cover.

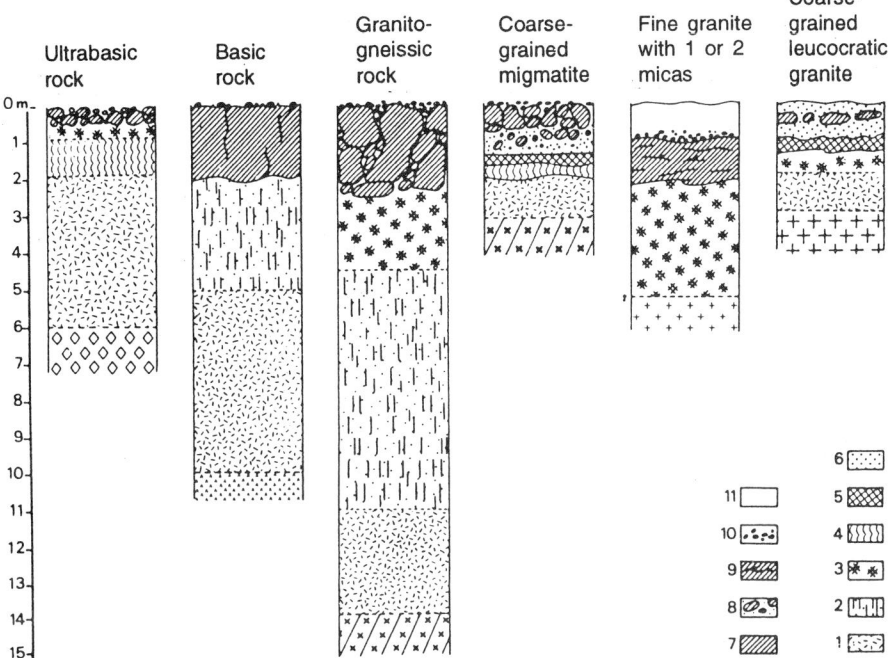

Fig. 42. *Relations between cuirasse profile, its degradation and parent rock (after Leprun, 1979).*

1—alteration ('green'); 2—variegated clay; 3—spotted clay; 4—greenish horizon; 5—illuvial horizon; 6—eluvial horizon; 7—cuirasse; 8—eroded cuirasse; 9—horizontally fissured cuirasse; 10—nodules at surface; 11—covering soil.

2.6.1.5.2 *Fossilised paleocuirasses under sedimentary cover.* The '**siderolitique**' which outcrops at the periphery or in small basins of the Massif Central is of Eocene age. It is mainly characterised by the abundance of red or variegated materials rich in ferruginous **pisolites**, wherein **kaolinite** is the only clay mineral. These siderolitic materials are locally indurated to ferruginous cuirasses.

Based on the studies of Daugas (1981) in Perigord, a paleoalterite with a cuirasse of this type is schematised in Fig. 44. The Toarcian (L. Jurassic) argillites formed at the base of the profile of a melange of illite and interstratified 10–14 sm were progressively and completely replaced by kaolinite and upwards became well crystallised towards the ferruginous facies. In the cuirasse only a few grains of weathered quartz subsist (Fig. 45) which possibly disappear locally. The crust is constituted of *goethite and aluminous haematite* only. Kaolinic laterite, overlain by a cuirasse as described by Leprun (1979) is found here. It is possible that the cuirasse could have developed within the profile, while the subjacent horizons were already present, but the **silicification** which affects the top of the profile must be taken as the result of very late alteration.

In the Apt Basin the Albo-Cenomanian deposits are glauconite-bearing marine sandstones. During an emergence in the Upper Cretaceous part of these green sandstones underwent **meteoric alteration** consisting of **desilicification and kaolinisation**, resulting in the formation of almost red ochreous sands (Triat, 1979). Towards the east of the basin, where the alterite is relatively thick, the profiles are often observed grading from

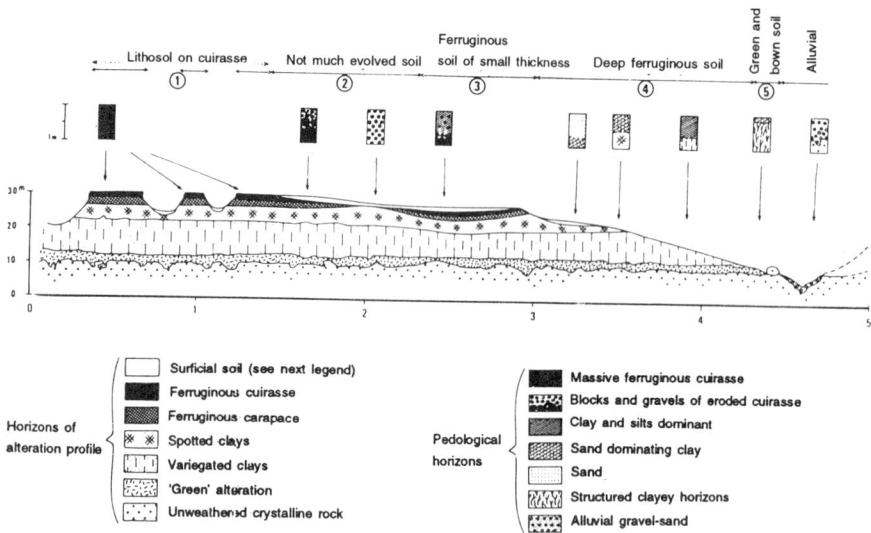

Fig. 43. *Synthetic scheme of an interfluve in crystalline region of Saheliax climate, West Africa. Distribution of main soil types according to truncation of alteration horizons (after Zeegers and Leprun, 1979).*

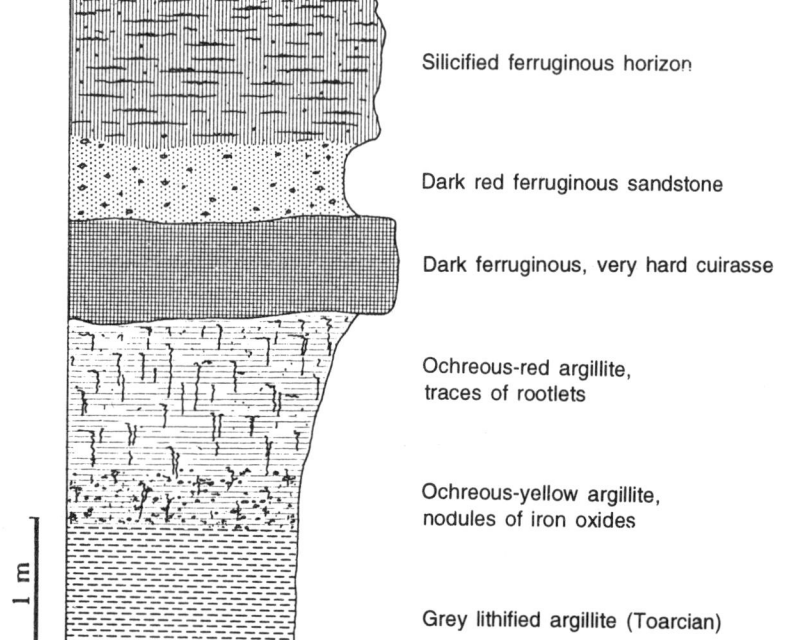

Fig. 44. *Ferruginous paleocuirasse in 'siderolitique' of Perigord. This alteration profile was developed on Toarcian argillite (after Daugas, 1981).*

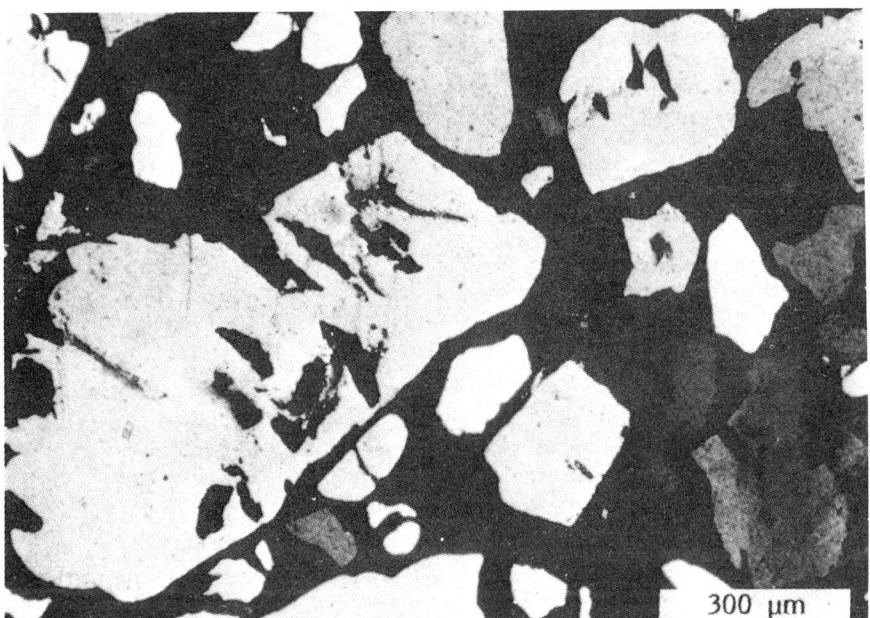

Fig. 45. *Corroded grains of quartz in an opaque ferruginous matrix. Ferruginous cuirasse in 'siderolitique' of Perigord. Microscopic observation under natural light. Sample, F. Daugas.*

glauconite-bearing sands to **ferruginous flagstone** (Fig. 46, after Guendon and Parron, 1983). The green sands progressively become whitish-yellow, then yellowish-orange but retain their structural and sedimentary features. Glauconite as well as smectite disappear progressively towards the top and kaolinite appears and becomes the exclusive mineral from the time of appearance of yellow sands. Goethite, which causes **rubefaction**, accumulates in a ferruginous cuirasse, in which **leaching** and **accumulations** obliterate primary structures in preference to secondary nodulisation. The ferruginous and kaolinic sandstone at the summit of the profile seems to be a part of the paleoalterite. The assemblage is sealed by Eocene sediments. It is to be noted that elsewhere in the same epoch **bauxitic** formation developed on more southern argillo-carbonate parent rocks. Similar structures in iron facies are found in the profile less affected by karstic withdrawal. This last ferruginous cuirasse is often rich in Al (Laville, pers. comm.).

2.6.1.6 Diagnosis and principles of interpretation
Horizons which have undergone redistribution or accumulation of iron oxide are readily identifiable by the naked eye because of their colour. The greatest difficulty is to *date the redistribution* at the outcrop. It is necessary to be on guard against confusing ferruginous concentrations developed in Quaternary climates with more ancient alterations.

Identification of mineralogical phases related to Fe^{3+} is, contrarily, quite difficult in natural facies, which are generally composite: the mixture of goethite and haematite may be identified by X-ray diffraction if the angles of diffraction used by the operator are large (Schwertmann and Taylor, 1977). The *degree of* **substitution** *of Al in goethite and*

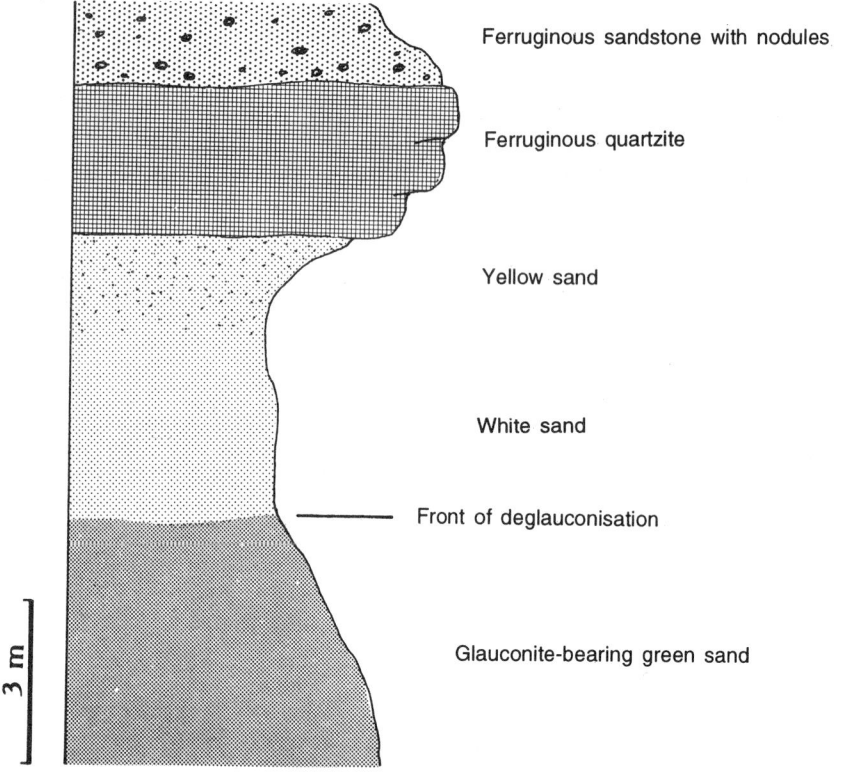

Fig. 46. *Paleoprofile of alteration indurated by goethite to form cuirasse of the Apt Basin, Albo-Cenomanian, Rustel region (after Guendon and Parron, 1983).*

haematite may also be estimated by analysing displacement of peaks on X-ray diffractograms (Périnet, 1974) but the most precise method for studying the substitutions is by Mössbauer spectroscopy (Lamouroux et al., 1977).

The red or ochreous colour due to iron oxides is a good index of **emergence**. Apart from certain argillite layers, marine formations which have undergone diagenesis in marine environments are almost never red. It is mainly the hydric paleoregime of the depositional environments which can be determined precisely by facies analysis:

— *Hydromorphy* related to water table *fluctuations:* colours change in a perpendicular direction to the principal plane of sedimentation.

— *Hydromorphy related to paleosols:* marmorisation takes place.

— *Hydromorphy in warm climate*, indicated by the presence of plinthite or petroplinthite.

— *Well-drained* environment in a relatively *warm* and *humid* climate: formation of cuirasse;

— *Oxidising environment*, consequently well drained with little biological activity: rubefaction; first stage leads to formation of red beds.

EXAMPLES OF PALEOALTERITES AND PALEOSOLS 67

─────────────── FERRICRETES ───────────────

Morphology:
Almost continuous horizons; very hard cuirasse overlying less-resistant carapace.

Facies:
Massive, nodular, pisolitic, scoriaceous, fine gravelly; colour, red to rust.

Mineralogy:
Aluminous goethite and haematite, kaolinite.

Substrata:
Endogenous or diverse sedimentary, generally parent rock.

Frequency:
Demarcates certain paleosurfaces. Often associated with siderolitic and bauxitic facies.

Paleoenvironment:
Warm and humid climate, long duration of alteration.

Convergence:
With sedimentary rocks containing reduced iron which are generally oxidised at the outcrop.

2.6.2 Aluminous Paleoalterites

Under conditions of meteoric weathering, aluminium is one of the least mobile elements and therefore likely to concentrate in some special types of alterites in two distinct modes:

— *Associated with silica*, it leads to the neoformation of **kaolinite** in the profile, or *associated with iron* gives rise to **aluminous haematite** (up to 15% Al) and/or **aluminous goethite** (up to 33% Al);

— *Dissociated from other cations*, it gives rise to **gibbsite** (oxyhydroxide of Al) or **boehmite** and **diaspore**. Gibbsite may occur in alterites formed in temperate climates (Tardy, 1969) or in certain **podzols**, but is relatively unstable and less abundant. Only warm, humid and highly hydrolysing environments produce an abundance of hydroxides of Al, economic concentrations of which are manifested as *bauxites*.

2.6.2.1 Recent and present-day laterites

The term **laterite** has been given diverse meanings by different authors. The general definition given by Millot (1964, p. 135) is adopted here: *'ensemble of weathering products of the intertropical zone'*. If aluminous accumulations are found in laterite, it must be remembered that they mostly occur in pairs with accumulations dominant in iron; the term generally used by pedologists to characterise this type of evolution is **ferrallitisation**. It is to be noted that the numerous laterites dispersed over the surface of continents are not in equilibrium with the climate of the region in which they occur, which proves that they are ancient and more or less fossilised. The processes have been modelled at places (Becker, 1992).

An instance of subrecent *laterisation* in Hawaii was studied and dated (Valeton, 1972, p. 88). Weathering profiles 30 to 50 m thick developed on basaltic flows 10,000 yr old. The prevailing topography precludes lateral transport. Kaolinisation is common throughout the landscape. Gibbsite takes precedence over kaolinite wherever percolation of water is most active. This alteration is exceptionally rapid because the conditions are practically ideal: the basaltic parent rock is very poor in silica and very porous so the resultant laterite remains highly porous. The annual rainfall is several metres and the temperature high.

Earlier studies of laterites in Equatorial Africa (Millot, 1964) and the more recent study of laterites of the Ivory Coast (Boulangé and Bocquier, 1983) reveal that the formation of alumino-ferruginous cuirasse implies the following mechanisms:

— *Weathering which eliminates soluble cations* but fully preserves the structures of the parent rock.

— *Transformation which obliterates these structures,* when the alterite is intruded by gibbsite, haematite and goethite.

— A complex succession of *internal readjustments* leading to nodular facies, followed by pisolitic facies (Fig. 47).

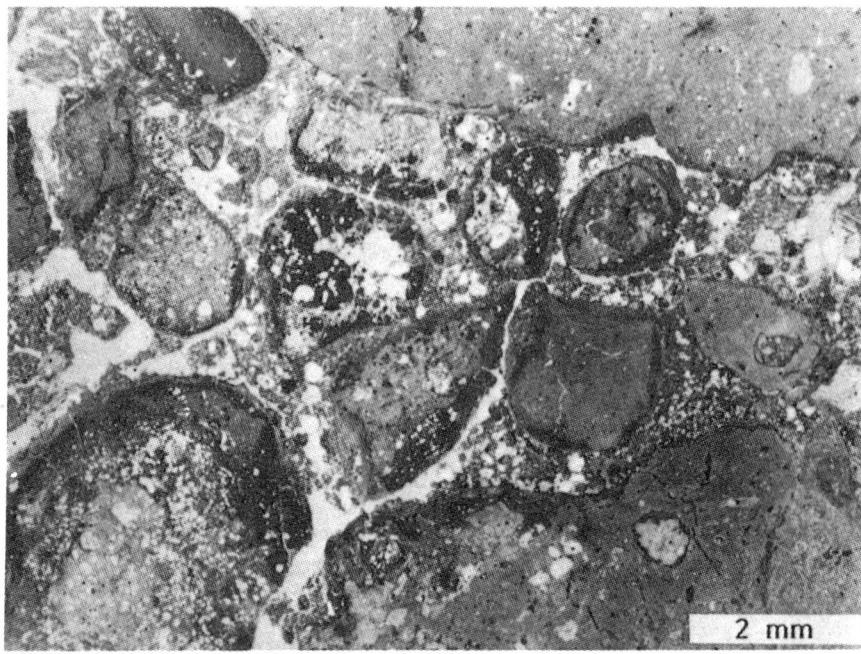

Fig. 47. *Lateritic bauxite (Ivory Coast). Enveloping of elements has given rise to frequent pisolitic structures in these formations.*

Whatever the parent rock or the climatic conditions it seems that the laterisation occurring due to simple geochemical alteration is not capable of concentrating aluminium; biological activity plays an important role (Erhart, 1967). Some ions, such as Na^+, and Ca^{2+}, are readily removed from the profile but not **silica** except when plants intervene. Lovering and Engel (1967) demonstrated that a vegetal cover of common herbal plants is capable of 'extracting' approximately 860 kg of **silica** per hectare per year from crystalline rocks. This silica is transformed into composite organo-minerals or **opal**; falling on soil, it is readily rendered soluble and mobilised in profiles. It is estimated that other conditions remaining unchanged, plants with shallow roots are capable of extracting the entire silica from indigenous rocks, including acidic rocks, up to a depth of 10 cm in 2000 yr. This data cannot be transposed to equatorial forests where weathering profiles are often more than tens of metres thick. However, it provides useful information for geological interpretation.

2.6.2.2 Fossilised laterites in ancient sequences

Millot (1964, p. 170) proposed many examples of **'siderolitique' formations and interpreted them as fossil laterites**. The examples chosen from Perigord or the Cretaceous of Provence to discuss the formation of ferruginous cuirasses are *de facto* fossil laterites (Sec. 2.4.1.5) in which iron has preferentially accumulated. Aluminium concentrations occur as pairs with iron but these are often discrete.

The example that follows is from the Middle East, mainly based on the works of Goldbery (1979, 1982) in Israel. He observed Lower Jurassic beds discordantly overlying the Triassic in the desert of Negev. These Jurassic sediments encompass an ensemble of facies derived from lateritic alterations. A schematic model is proposed in Fig. 48, integrating all the observed diverse facies. The chronology of events could have been as follows:

Stage a: A marine regression caused the *emergence* of carbonated and sulfurated Triassic rocks. Alterations which followed gave rise to a karstic relief in these Triassic rocks. A fluvial network in the region transported **allochthonous** remnants derived from *ferrallitic alterites:* these are the ferruginous pisolites which accumulated locally as conglomerates and as immature *'lateritic arenite'* consisting of ferruginous and kaolinic debris embedded in a clay matrix (kaolinite). These minerals underwent diverse modifications in the alluvial plain. *Soils appeared.* A few of these facies fossilised *in situ* in non-karst regions; horizon B of argillaceous accumulations, **slickensides** and traces of deformation (pseudoanticlines) are probably related to alternation of weathering and drying providing a basis for supposing that the surface of the soil had been affected by undulations of decimetric scale (gilgai relief). Carbonated or gypseous concretions may have appeared at the base of the profile; such paleosols are similar to present-day **vertisols**, i.e., ferrallitisation did not take place on the alluvial plain since these soils were not enriched in Fe or Al.

Stage b: A transgressive episode caused a radical change in the environment: the supersaline lagoons became the seat of *deposition of carbonated and marly sediments.*

Stage c: Fluctuations in sea level abetted alteration of allochthonous lateritic materials. This was particularly active in *karstic regions* where the conglomerates and lateritic arenites are leached out, especially part of the iron, which means that the percolating waters were *reducing* in nature. Locally this gave rise to accumulation of a particular kind of kaolinic clay known as **'flint clay'**. These flint clays are hard, non-plastic, with conchoidal fracture. They are rich in alumina; minerals of the kaolinite family are accompanied by a little **boehmite** and **diaspore**.

The phases of alteration envisaged encompass a vast ensemble: a series of paleoalterites correctly dated as Triassic-Jurassic have been identified in the Middle East (Abed, 1979; Goldbery and Beyth, 1984). This includes **red beds** to paleoalterites, eventually transformed into bauxite or **alunite** deposits.

This example reveals the complexity of the phenomena and facies related to oxidised paleoalterites. Already highly altered matrices in the upstream zone are deposited in the area of sedimentation. Evolution follows deposition, discretely in the paleosols and *much more vigorously across the karstic network* when there is better **drainage**. Partial **leaching** of iron is associated with percolation of reducing solutions, which acquired this property

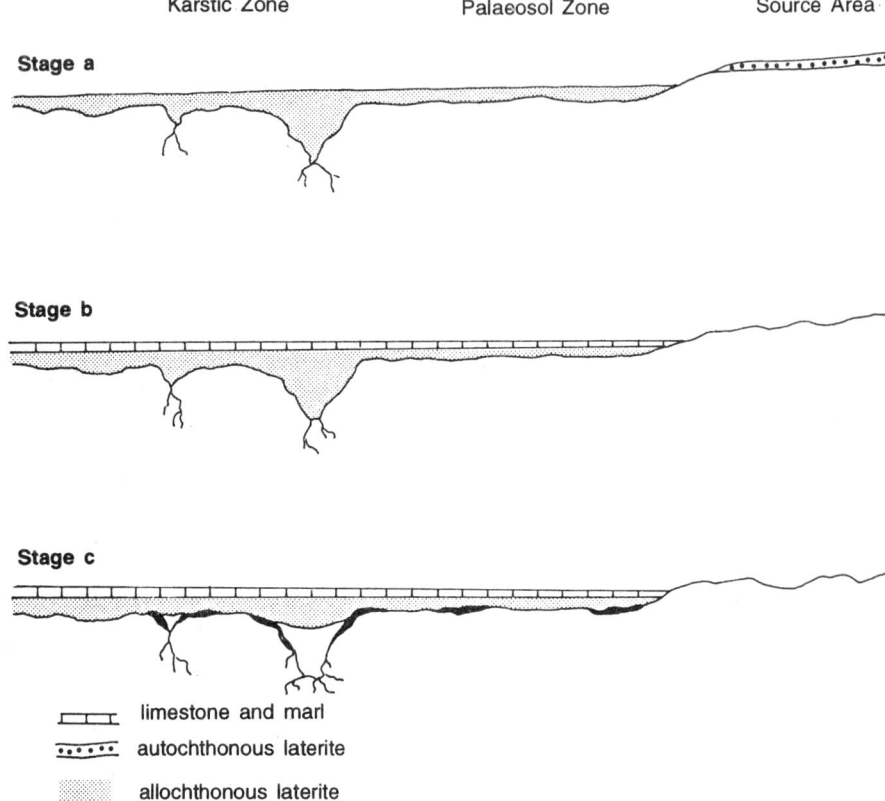

Fig. 48. *Stages of alteration of the Lower Jurassic at Negev (after Goldbery, 1979).*

Stage a: Allochthonous lateritic deposits cover the region.
Stage b: Rapid marine transgressive deposits of limestone and marl.
Stage c: Meteoric alteration recommences following a marine regression; leaching transforms part of allochthonous laterite into flint clay.

from the sediments subjacent to the alterites. In this case, therefore, the sedimentary cover, probably rich in organic matter, directed the last phase of alteration. This example is identical to the 'flint' mined at Ollières in Var (Esterelle, 1967; Lajoinie and Laville, 1979; Laville, 1981).

2.6.2.3 Bauxites

Rocks containing more than 40% Al_2O_3 and less than 15% SiO_2 are conventionally defined as bauxite. A *threshold of around 15% SiO_2* separates the dominant aluminium hydroxides facies from the preponderant aluminium-silicate facies. It is certain that in the fossil laterites described earlier the content of aluminium hydroxides is generally less than 40%; the content of Al_2O_3 is considered low compared to bauxite. The fact is universally accepted, especially in '**siderolitique**'. An inventory of Cretaceous bauxitic formations in Provence and in Languedoc (Lajoinie and Laville, 1979, 1980) led to comparable conclusions: of the 301 km^2 of bauxitic formations, an area of 159 km^2 is mineralised, whereas only 37 km^2 is rich for economic exploitation. It is therefore evident that *very specific conditions are required for the formation of bauxites.*

Bauxites have been recognised from nearly *all geological periods* starting from the Cambrian; however, certain periods like the Cretaceous were particularly favourable for their formation (Valeton, 1972, p. 61). The Carboniferous witnessed the advent of massive development of bauxite formations. Kulbicki and Vetter (1955) described bauxite facies at the east periphery of Decazeville basin. They are overlayed by productive Stephanian beds. They are red, argillaceous, often pisolitic, rich in kaolinite but also contain boehmite and gibbsite. The following examples represent deposits of economic interest.

2.6.2.3.1 *Bauxitic profiles on paleokarst.* A number of bauxite beds occur on karstic carbonated substratum in Midi France. These beds have a complex history, involving many cycles of alteration-erosion-sedimentation (Combes, 1984). Guendon and Parron (1982, 1983) succeeded in identifying karstic pockets in the Alpilles hills in which definite synchronisation occurred between the mechanisms which sculptured karstic morphology and those which controlled bauxitic alteration.

Essentially two mechanisms were involved in their evolution:

— Dissolution of the karst under the bauxitic cover, resulting in a downward pull and consequently a *mechanical sinking* of the bauxite cover.

— Alteration of bauxitic cover with the appearance of a veritable lateritic profile; this evolution involved *geochemical sinking* of bauxite horizons.

The interaction between these two phenomena gave rise to various types of beds:

— Beds with **fragmented and displaced relict structures,** as a result of mechanical driving in, with individualisation of pseudoblocks and pseudopebbles.

— Beds with **successive superposed profiles** wherein progressive mechanical driving in of the entire lateritic profile on the material took place. The material underwent the primary stage of alteration (Fig. 49).

— Beds with total **superimposition**, characterised by a very thick unique profile, which is superposed on the earlier profiles; this evolutionary stage was induced by the preceding one in a karstic system in which a geochemical sink predominates over a mechanical sink.

These sediment types represent different stages of bauxitic evolution in the same karstic landscape.

2.6.2.3.2 *Bauxitic paleogeography.* Combes (1978a) proposed a paleogeographic model to explain the geometry of bauxite beds of the third horizon of Parnasse zone in Greece. This bauxitic horizon belongs to the end of the Lower Cretaceous and rests on karstic Albo-Aptian limestones; several observations have helped to correlate the geometry of the bauxitic bed with the *paleogeography of subjacent formations.* The bauxitic formation is trapped in a primary **karst** whose morphology varies between three principal types (Fig. 50a).

— *Type 1:* Strong karstification: beds are elongated, major axis measures 150 to 2000 m, minor axis 50 to 200 m; maximal thickness around 15 m.

— *Type 2:* Strong karstification: beds occur as pockets of 15 to 30 m in diameter, more or less braided; the pockets are 10 to 25 m deep, with an irregular bottom surface.

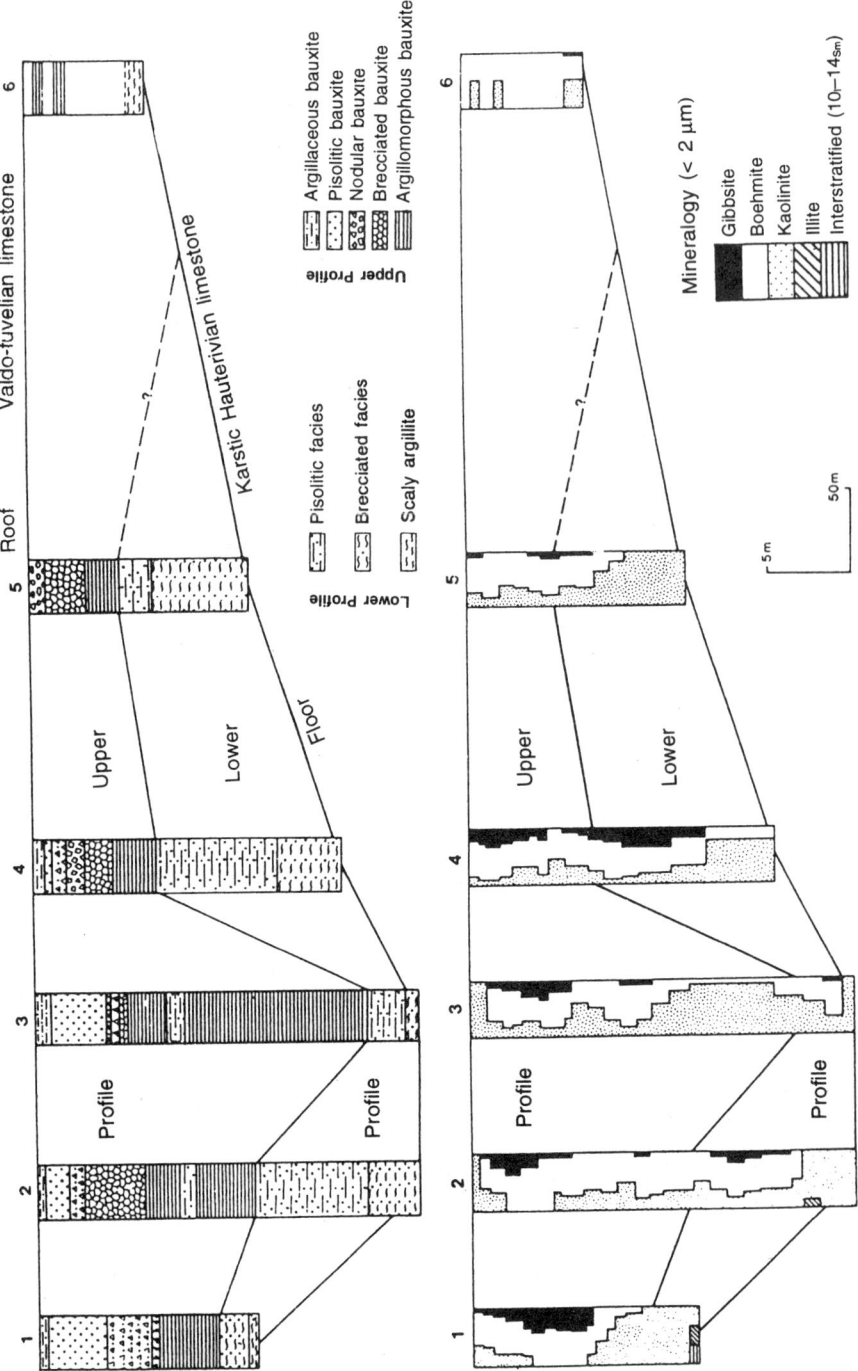

Fig. 49. Superposed bauxitic profiles in sediments of the Alpilles (after Guendon and Parron, 1983). Toposequential suite along a karstic depression; petrographic facies are represented in the top diagram and mineralogy of fine fractions shown in the bottom diagram.

— *Type 3:* Weak karstification: beds may differ in shape, are normally 3 to 4 m in thickness, the bottom surface covered with small decimetric mounds.

The distribution of these diverse types of beds is not accidental but a function of the subjacent formation (Fig. 50b and c): it is a limestone formed on the floor of the inner basin in which feeble energy zones have been recognised with essentially micritic facies (inner basin), and the border zones with oolitic, oncolitic and more or less dolomitised facies indicative of high energy and influence of continental waters. During the emergence that follows deposition, karstification develops preferentially at the border and slope of the basin; it may be due to the relief of the paleotopography or to more favourable petrographic facies. Whichever be the case, emplacement of the bauxitic beds extends over the karstic morphology and hence over the **prebauxitic paleogeography**. This is a method of prospecting that can be transposed to other beds.

2.6.2.3.3 *Bauxitisation*. The economic importance of bauxite has motivated numerous studies, all of which cannot be detailed in this book. Cases which have been very intensively studied, however, permit formulation of some principles for understanding the genetic environments of these complex rocks.

In the examples mentioned from the Middle East and Europe, the bauxite rests on a carbonated karstic base. Yet instances are known of bauxite overlying an alumino-silicate substratum from which it was derived almost directly. Bardossy (1981) calculated that in Europe 92% of the bauxite deposits belong to beds of the karstic type but considering all the global deposits, only 14% belong to this category, the remaining 86% overlying alumino-silicate substrata. It is in fact *an accumulator* because the **dissolution of carbonates** is maintained under an alterite cover. However, in agreement with Laville (1981) it may be noted that these two types of deposits are not opposed, but rather manifestations of the same phenomenon. The primary condition is that the substrata have a good drainage system which can influence the percolation and subsequent weathering of the alterite into bauxite. From the chemical composition point of view, the *bauxitic formation* generally consists of a vertical sequence of three members (Laville, 1981); such an ideal sequence may be repetitive or terminated at one or more facies; the following sequence is recognised from top to bottom:

— *Upper siliceous facies* rich in kaolinite, in which the content of titanium minerals (anatase, rutile) are always more than 1% but rarely 6%; occasionally truncated by an erosion linked to subjacent deposits; it often bears traces of hydromorphy consecutive to submergence of the bauxite formation, which is gradually buried under mud more or less rich in organic matter.

— *A middle bauxite facies*, economically exploitable, as it is rich in Al-hydroxides.

— *Lower siliceous facies*, rich in kaolinite.

The disposition shown in Fig. 49 implies an *in-situ* evolution of alterite; in the bauxite beds alteration is always **autochthonous**, even if the material that undergoes weathering is **allochthonous**. It is therefore possible to propose a chronology of events resulting in the formation of real bauxites.

— *Vigorous meteoric alteration*, which on rocks containing alumino-silicates, eliminates **alkalies, lime** and **silica**. It presupposes a very hot and very humid climate, good **drainage** and long duration.

— Weathering may affect the same material at different sites as a function of *successive reworking* but the aluminium hydroxides are always fragile and those found in the beds must have formed *in situ* or in the immediate environment.

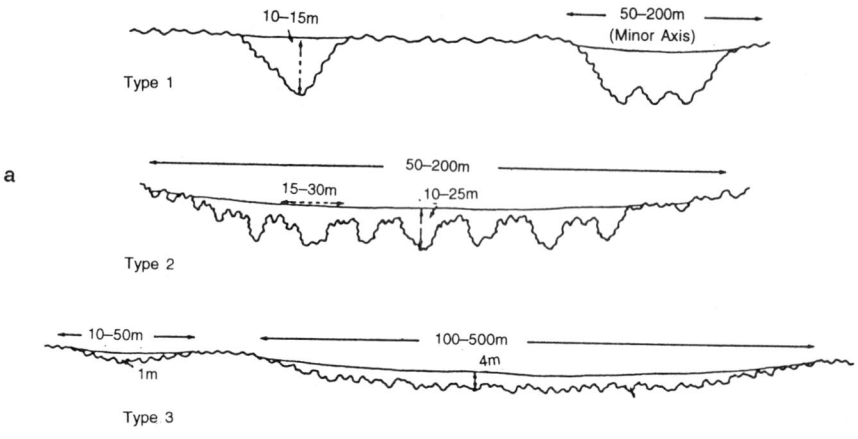

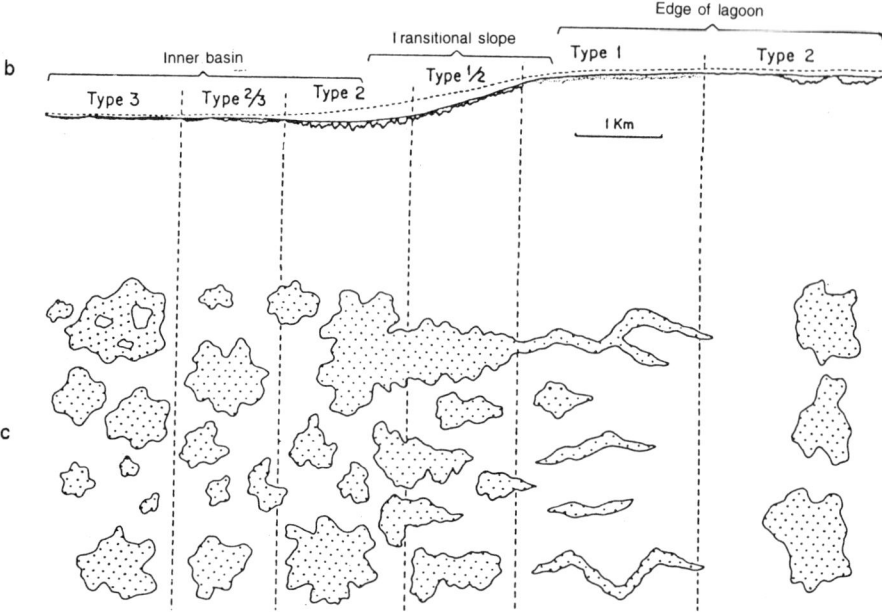

Fig. 50. *Relation between paleogeography and metallogeny of bauxite (Third Horizon, Parnasse Zone, Greece; after Combes, 1978a).*

a) Transverse section showing karstic morphology of three main types of sediments. Indicated dimensions give order of the size.

b) General section showing distribution of sediments in relation to the paleogeography of the subjacent limestone. Dotted line indicates the presumed paleotopography at the time of emergence and before karstification; it permits appreciation of the importance of emergence and karstification (exaggerated scale).

c) Plan view of the relative disposition of mineralised bodies.

— Since the alterite is sealed by recent sediments, their nature controlled diagenetic transformations; the lignitous and sulfide layers often present at the top of bauxites may become oxidised and liberate *sulfuric acid*, which *removes iron from the bauxite* and eventually precipitates some **alunites**—all of which improves the quality of the deposit; contrarily, if the percolating water contains dissolved silica, *aggradation of kaolinite* to the detriment of aluminium hydroxide is favoured. This then is the origin of the upper siliceous facies, formation of which reduces the importance of the deposit.

This polyphased evolution is sufficiently selective so that the true bauxite facies can be clearly distinguished from the surrounding bauxitic facies: in Midi France the average silica content in bauxite is 8%, while in the upper and lower siliceous facies it exceeds 20% (Lajoinie and Laville, 1979).

2.6.2.3.4 *Diagnosis and principles of interpretation.* Even if **pisolitic** or **gravelly** facies coloured by iron oxides are sufficiently characteristic, it must be accepted that lateritic profiles do not present an organisation which is precise and regular. Even after eventual reworking and diagenetic action, the principal criteria for diagnosing fossil laterites are finally of the *geochemical* and *mineralogical* order. The more the content of Al_2O_3 in the rock, the more the lateritic tendency of the profile. Kaolinite, like all hydroxides of aluminium, is identified correctly by the X-ray diffraction method.

A positive Al anomaly, or more appropriately, the presence of appreciable quantities of aluminium hydroxides in an ancient sequence is an *excellent criterion for identifying a* **paleosurface** in which weathering has been taken to the extreme, either because the climate was particularly warm and humid or the break in sedimentation was long enough to allow complete alteration. In very ancient sequences (Sec. 2.8.2) it appears that this type of alteration was too rare to be identified.

Bauxites cannot develop except by the conjunction of *very specific conditions*, which explains why they constitute a very small proportion of fossil laterites.

2.6.3 Oxides and Hydroxides of Manganese

Under surficial conditions manganese is soluble in the Mn^{2+} state and precipitates as oxides of Mn^{3+} or Mn^{4+}. The degree of oxidation passes easily from 3 to 4 or vice-versa, which results in a *complex mineralogical melange* (McKenzie, 1977). In standard works geologists identify the precipitates of manganese macroscopically as *black-coloured dendrites*, almost continuous pellicules (mangans), even as small nodules; all these deposits are opaque under the microscope in transmitted light. Mn^{2+} exhibits a comparable behaviour as Fe^{2+} but oxidation of Mn^{2+} requires higher values of **Eh** and **pH** (Guillet and Souchier, 1979). Mn is often rendered soluble under very slightly reducing conditions and in contrast to iron, migrates through fissures towards the surface where it is oxidised in contact with air and precipitates. This mechanism is effective in sediments or aerated soils subjected to **short hydromorphic periods**, perhaps even days (Veneman, Vepraskas and Bouma, 1976). The climatic conditions are equally important; in cold climates manganese forms complexes with the organic matter and is eliminated from the profile, while in warm and humid climates it may accumulate at the bottom of the profile (Roy, 1981). In ancient continental sequences manganese deposits occur in diverse forms. Some examples are given below.

In the Oligo-Miocene molasse of Aquitaine certain layers of clayey-sand with polyhedral structures contain fissures lined with black pellicules a few centimetres or tens of centimetres in dimension. This coloration developed horizontally and extended laterally for tens of metres. The MnO content, which is generally less than 0.1% in the rock, attains a level of more than 0.5% in these layers. These deposits indicate a very temporary hydromorphy above the water table. Certain rubefied B_t horizons may have been affected by the phenomenon, showing that clay mineral accumulation stops when temporary phases of obstruction are induced in the surficial horizons. It appears that in very ancient sequences these traces are not as well fossilised as those left by iron hydroxides.

In the same molasse of Aquitaine certain carbonated crusts of pedologic origin contain manganese coating fossilised by the sparite in tubules and fissures (Fig. 51). All evidence leads to the belief that Mn goes into solution in relatively acidic surficial pedologic horizons. It infiltrates up to the limestone horizon where the *high pH value of the interstitial water facilitates its precipitation* (Meyer and Guillet, 1980). Van Straaten (1978), on studying the dendrites in limestones of Solenhofen, arrived at a similar conclusion. Dendrites lining the fracture surfaces very often indicate recent alteration but when they are fully enveloped in a lithified body of limestone, they indicate the percolation of meteoric waters in the environment of genesis (Crouzel and Meyer, 1977a).

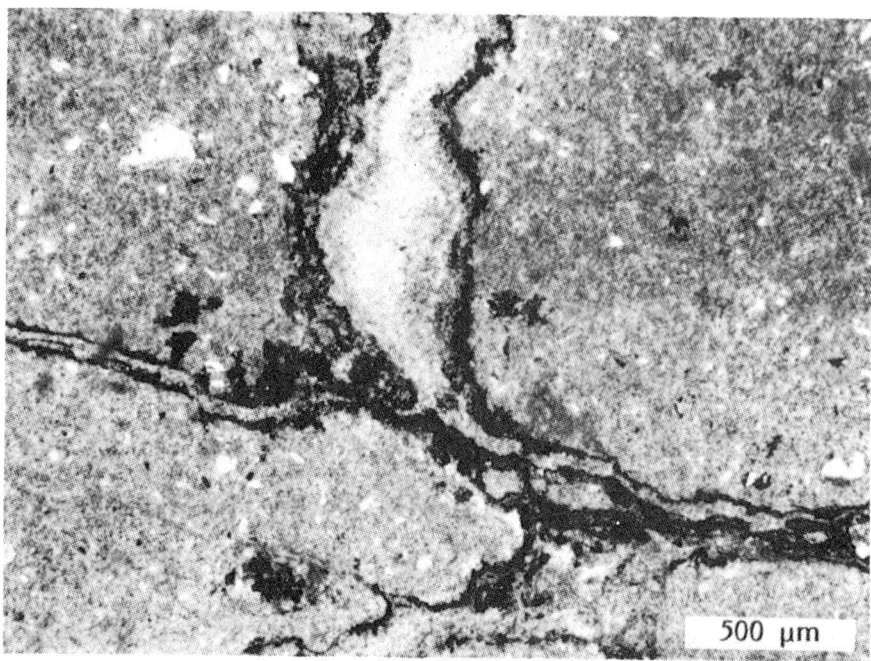

Fig. 51. *Coatings of manganese oxide in limestone crusts of Aquitaine (Miocene). Black manganese oxide lines the fissures which are subsequently filled by sparite.*

In warm and humid climates these surficial migrations may result in considerable accumulation of Mn (Roy, 1981); the conditions favouring this process are listed below:
— a *very long period of weathering*, capable of extracting Mn from primary minerals;
— formation of a *thick alteration blanket*,
— *less pronounced relief* which reduces the erosion.

EXAMPLES OF PALEOALTERITES AND PAEOSOLS

FOSSIL LATERITES

Morphology:
More or less stratiform.

Facies:
Massive, nodular, pisolitic; red or rust in colour due to oxides or hydroxides of iron.

Mineralogy:
Hydroxides of aluminium and iron, kaolinite.

Substrata:
Very permeable; often weathered parent rock containing alumino-silicates.

Stratigraphic distribution:
Certain preferred periods, which vary from region to region globally.

Paleoenvironment:
Warm and humid climate, long periods of alteration.

Two opposing diagenetic evolutions are possible:
— A resilicification of hydroxide of Al when subjacent sequences liberated silica in solutions (for example, from sandstones).
— A deferrification resulting in bauxite when subjacent sequences rich in organic matter liberated reducing and acidic solutions.

These accumulations do not acquire an economic value unless the parent rock itself is rich in Mn (for example, rhodocrosite).

In conclusion:
— Coatings of manganese oxides are identified by their black coloration.
— In paleosurfaces they indicate phenomena of *emergence* and *temporary hydromorphy*.
— They are often too common to authorise a complete interpretation by themselves.

2.7 Paleosols on Volcanic Rocks

2.7.1 Present-day Soils on Volcanic Material

Development of soils on *basic rocks* (basalts) is much *more rapid* than on acidic rocks (trachytes). Moreover, pedogenesis is more rapid on volcanic *ash* because of the microdivision of particles than on unfractured massive rocks although it is essentially vitreous. The normal pedogenesis on these rocks is *andolisation* when the climate is sufficiently humid. **Andosols** are not much evolved soils and differentiate into discrete horizons. They are generally dark in colour and have specific mineralogical characteristics:
— They are rich in **allophane** and **imogolite** (poorly crystallised hydrated aluminosilicate; Wada 1977); they may evolve into minerals of the kaolinite family (metahalloysite and interstratified 7–14 sm).
— Phosphate ions are retained as aluminium phosphates.
— Liberated silica may be fixed in the soil as **silans**.

If the climate permits long and repeated phases of desiccation, amorphous products evolve in an irreversible manner (Duchaufour, 1977, p. 210) and andosol is transformed into other soil types: **andic, andopodzolic, fersiallitic** etc. **soils**.

2.7.2 Examples of Paleosols on Volcanic Material

The most discussed examples are naturally found in the Recent, Quaternary and Tertiary sequences. Because of the repetitive nature of volcanic eruptions, many paleosols on volcanic rocks are fossilised beneath a flow or deposit of ash, which destroys all vegetation.

The paleosols on **Miocene basalt** in the Massif Central of France reveal this phenomenon (Dejou, Chesworth and Larroque, 1982; Chesworth et al., 1983). Certain sites show up to six paleosols intercalated with basaltic flows. Depending on the time period separating two consecutive flows, as well as the intensity of erosion, the thickness of the profile varies from 1 to 3 metres. At the top of each profile thermometamorphism is quite systematic for a few decimetres, because the profiles are sealed by subsequent flows. Fig. 52 presents a type profile which has been interpreted as a paleosol of fersiallitic evolution. In addition to the global geochemical transformation, neoformation of minerals is evident: Initially **metahalloysite**, but quite often some **smectite** and **gibbsite** also form; iron liberated from the silicates is fixed as hydroxides and tends to accumulate. A coherent ensemble of transformations is established, permitting the supposition that the andosol stage was effectively bypassed to achieve a *fersiallitic differentiation* which was probably due to climatic contrasts of the epoch.

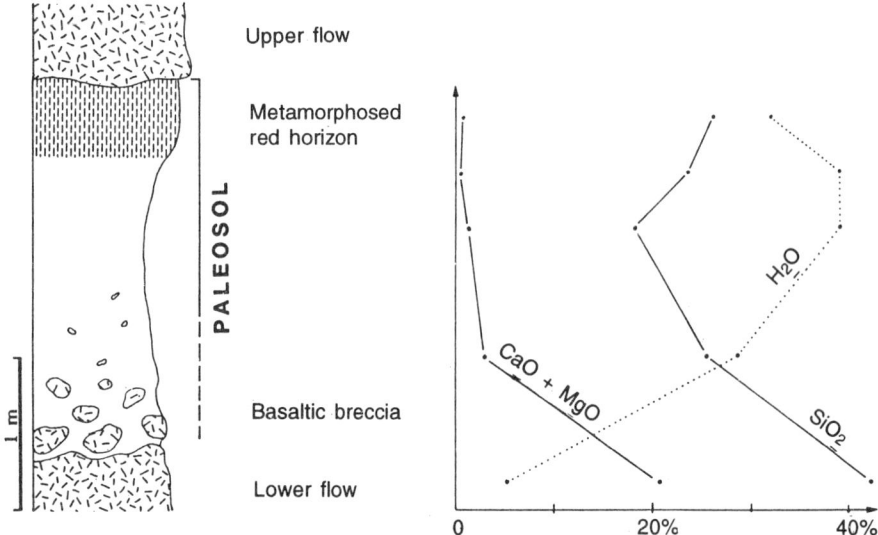

Fig. 52. *Fersiallitic paleosol on Miocene basaltic flows, Massif Central, France (after Chesworth et al., 1983). Right-hand graph represents a few geochemical evolutions of the whole rock: alteration is effected by the leaching of Ca, Mg and Si; steep rise in the percentage of water accompanies the neoformation of supergene minerals.*

In the Eocene volcano-sedimentary formations of Wyoming, Retallack (1981) identified a paleosol on a muddy flow of volcanic products in which silicified trunks of Sequoia are preserved in their original upright position. He distinguished in the field a **decolorised horizon A** and an **argillic horizon B** over a slightly altered horizon C. Silicification appears to have affected all the facies of the profile but laboratory results failed to confirm field observations.

In the more ancient volcano-sedimentary sequences, there is every reason to believe that meteoric alteration could also have left imprints. A very early stage is represented by the *lapilli of accretion*, a kind of small **pisolite** formed by agglutination of volcanic ash around rain-drops, also observed in the Permian of Esterel or of the Vosges. A case of great importance, though difficult, is that of **tonsteins**. These clear layers, varying in thickness from a few centimetres to a few metres, act as good stratigraphic markers in certain coal sequences and are generally considered to be of *cineritic origin* (Even, 1978). However, the evidence often favours meteoric transformations of the cinerite:

— Mineralogical composition of a tonstein may vary laterally in hectometres (G. Even, pers. comm.), which conforms with the variability of a paleolandscape.

— **Kaolinite** is generally abundant, which could have formed by the *recrystallisation of allophanes* which are highly unstable;

— Aluminous phosphates are abundant in certain tonsteins, in fact comparable to that found in andosols.

— Certain tonsteins could have resulted from the metamorphism of ashes due to the *spontaneous combustion* of carbonaceous products in the *environment of deposition*. The phenomenon of spontaneous combustion is observed in present-day pit-bogs of Senegal whenever a lowering of the water table takes place (Pezeril, pers. comm.).

2.7.3 Conclusions

— Andosols and andic soils, which are typical of the first stages of pedological evolution of volcanic materials, are *little differentiated* and consequently *difficult to identify as paleosols*.

— Unstable minerals such as allophanes are of no interest for the study of paleoalterites, but they may evolve into clay minerals (particularly the kaolinite family), which constitutes an index as good as *aluminous phosphates*.

— Investigation of paleosols in continental volcano-sedimentary formations is difficult, though possible, as evident from the investigation conducted in the Massif Central of France (Fig. 52).

2.8 Paleomantle of Alteration

Most of the paleoalterites described in the preceding sections present a succession of facies and an organisation of profiles which are comparable to present-day soils. In addition, a considerable volume of paleoalterites and paleomantle of alteration exists with less specific characteristics and hence are difficult to interpret. In these cases it is necessary to go beyond the simple pedological context and attempt an interpretation in the realm of geodynamics. The existing models that are used to provide an insight into their studies envisage external geodynamics and the genesis of *surficial formations*.

2.8.1 Examples of Paleoalterites on Hercynian Basement

2.8.1.1 Triassic basement-cover contact in south-western Vosges
Studies carried out in the open-casts between Plombières and Contrexéville (Krakenberger et al., 1980) have demonstrated that the migmatitic basement was covered by the Principal Conglomerate of small thickness which, in turn, was covered by a violet zone. Alteration in migmatite as well as in the sedimentary cover has been studied to decipher the phenomenon before and after its deposition.

The basement rock is arenised up to a thickness of one to two metres but the original structure and a certain cohesion have been maintained. Alteration essentially influences the plagioclase and **biotites** with **leaching** of Na and Ca. This alteration is thus taken care of by the simple **hydrolysis** of the least stable minerals. Investigations have been carried out in the Triassic cover on both the sandy matrix and pebbles from the basement. On the whole, the results are comparable between them and those obtained for the weathered basement. There is almost total leaching of the Na and Ca.

The argillaceous phase is dominated by **illite**, the crystallinity of which is utilised to estimate the degree of alteration (Fig. 53). The grade of crystallinity is slightly more pronounced in the sandy matrix and the pebbles from the basement than in the basement per se, but with no significant hiatus at the discordant surface.

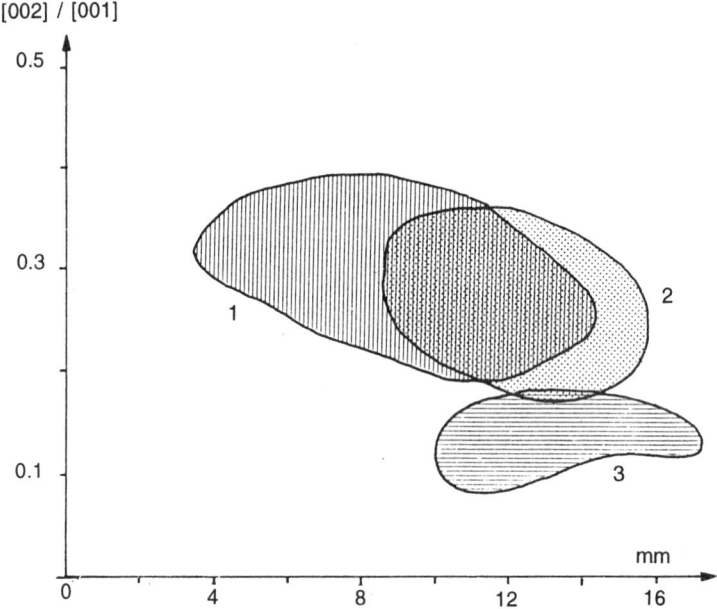

Fig. 53. *Esquevin diagram for three populations of* **illite** *(after Krakenberger et al., 1980).*

The mid-height breadth of the peak (001) is drawn along the abscissa; the ratio of the values of peaks (002) and (001) is plotted along the ordinate.
1. *illites from altered basement (13 data);*
2. *illites from sandy matrix of Triassic conglomerate (16 data);*
3. *illites from altered pebbles of basement (7 data).*

The greatest difficulty lies in *dating* these alteration processes. The pebbles from the basement are, for example, less coherent, even pulverulent, and they could not have been transported in this state; at least part of the weathering therefore occurred after deposition and could be related to burial diagenesis. However, evidence has proven that the basement was arenised before the first Triassic sediments were deposited:

— The base of the first sandy bed often shows casts (*sole marks*), which filled the small erosion structures grooved out of a basement that was relatively soft.

— Quartz in the sandy matrix is separated into two populations: 20% of the grains are angular, derived from the arenised basement, along with the worn-out grains of much earlier origin.

— The argillaceous phase of the sandy matrix is very similar to that of the weathered basement (Fig. 53); therefore, a local origin may be envisaged. However, this argument is ambiguous because burial diagenesis could have brought about a uniformity in the argillaceous assemblage.

The difficulty in interpreting such cases is inherent to them and recurs in comparable conditions. The mineralogy of the altered basement is relatively continuous with the subjacent sedimentary cover, a phenomenon which can be equally well explained by a local *sedimentary reworking*, or a *diagenetic evolution after deposition*.

2.8.1.2 Clay of basement and Triassic cover in Ardeche

In this region greenish-grey Triassic arkoses overlie Hercynian **granites** which are altered below the plane of discordance. These granites are very coherent, however, and contain less than 1% clay. The nature of these clays was studied from samples taken from four drill holes (Even and Samama, 1969). The results have been synthesised in Fig. 54 and provide some important facts:

— Clay minerals are *neoformed* in the *weathered basement*; they are not found at the base of the sedimentary cover, which *practically precludes* a **diagenetic neoformation**.

— Such an authigenesis indicates a *pre-Triassic continental weathering*; the differences from one borehole to another can be explained by the special nature of paleotopography.

— Almost complete profiles were removed by erosion and neoformed clays were dispersed into abundant illitic material, probably of a much earlier origin, and thus diluted.

— The argillaceous assemblage of the basement is often transformed at the discordant plane, influenced by the environment of sedimentation, or the milieu of diagenesis.

With respect to the methodology, such an example illustrates in particular that a paleoalterite could be preserved below the plane of discordance without a *diagenesis rendering the mineralogy of the assemblage and that of the cover uniform*.

2.8.1.3 Albitisation of basement beneath a Liassic cover in Rouergue

A study of diverse uraniferous beds in the south-western part of the Massif Central revealed that mineralisation is spatially located beneath the Triassic paleosurface, which was itself fossilised under the sediments generally dated to Lower Lias (Schmitt, 1983). Prior to uraniferous mineralisation, the sedimentary (Permo-Carboniferous) or metamorphic (orthogneisses, migmatites, mica schists) rocks beneath the paleosurface underwent weathering. This alteration was affected by albitisation and chloritisation descending to 100 to 200 m below the surface. Albitisation developed in different stages, well recognised at the hectometric scale, as also in the samples collected from the proximity of fractures. The principal stages of the sequence of albitisation are shown in Fig. 55:

— *Stage 1:* the fissures are intruded by *small pink crystals* of neoformed **albite**; this albite occurs mainly around potash-feldspars, oligoclase and micas.

— *Stage 2:* neoformed albite is present *throughout*; coloration is particularly evident around feldspars and in micas.

— *Stage 3:* entire rock is spotted with *pink zones* which coalesce into a pseudobreccia facies; the final stage occurs only locally and the rock may be composed of 50% albite and 50% quartz.

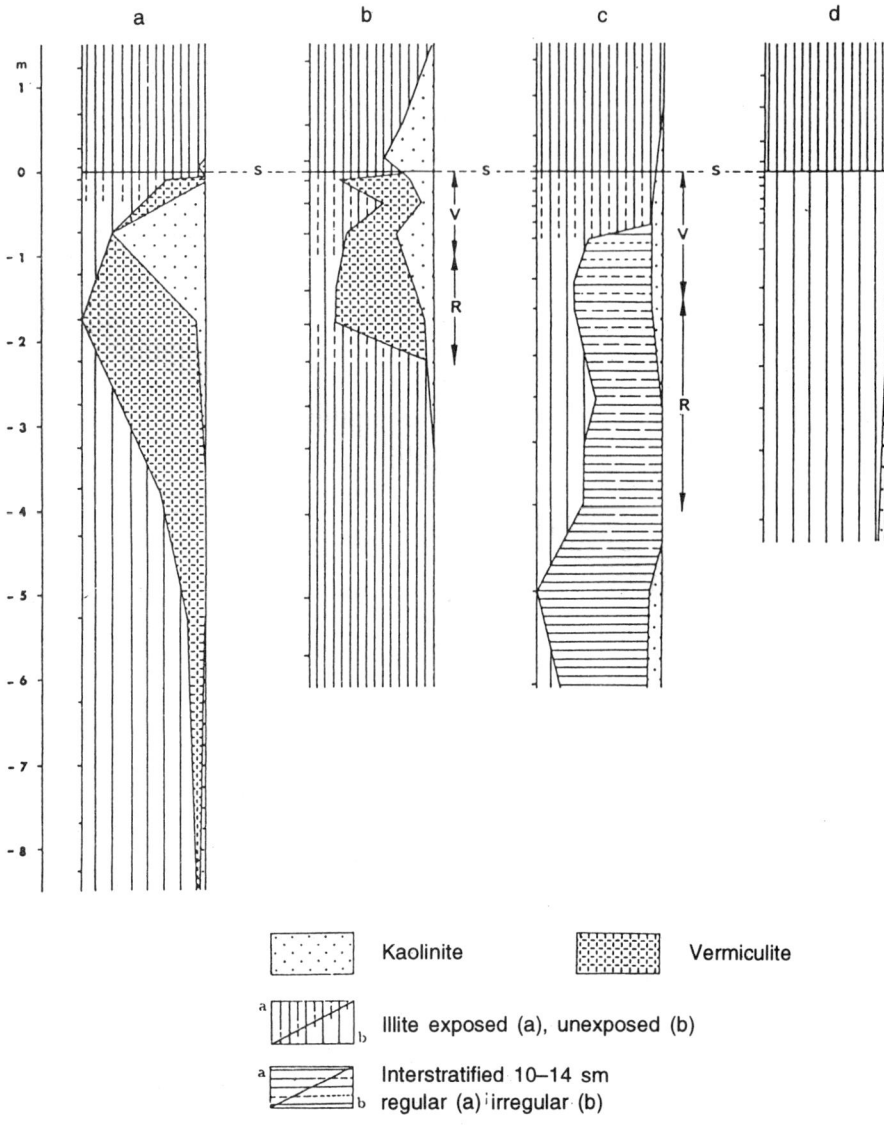

Fig. 54. *Profile of the evolution of argillaceous minerals at basement contact, Triassic, Ardeche (after Even and Samama, 1969).*

a, b, c, d correspond to four drill holes.
S: Discordant surface;
R: Red colour;
V: Green colour.

EXAMPLES OF PALEOALTERITES AND PALEOSOLS

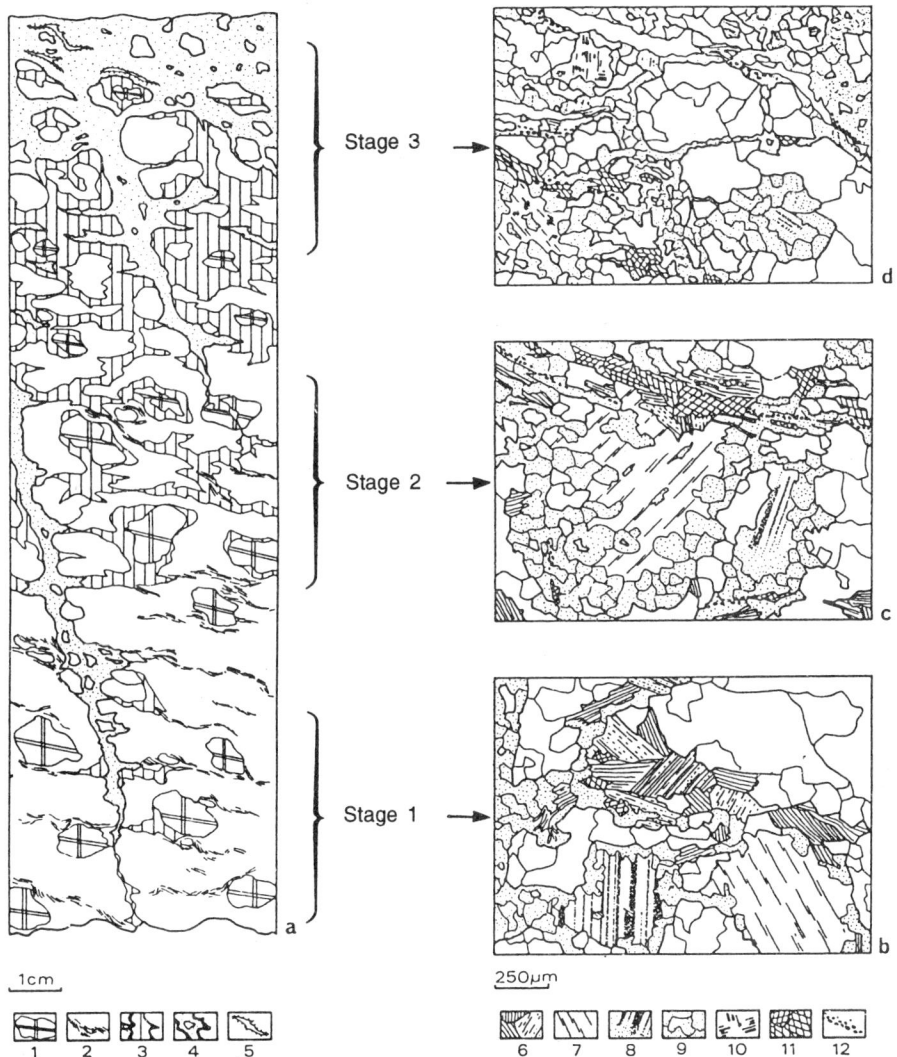

Fig. 55. *Sequence of albitisation beneath the pre-Triassic paleosurface. Observation in the orthogneiss of Rouergue (after Schmitt, 1983).*

a) macrosequence visible in a drill core; b, c, d) Schematised corresponding macrofacies.

1—potash feldspars; 2—mica; 3—rose-coloured zone; 4—rose-coloured zone with no specific structure; 5—void; 6—mica (mainly biotite); 7—potash feldspar; 8—oligoclase; 9—neoformed albite; 10—chequered albite; 11—siderite; 12—traces of leucoxene.

This type of paleoalterite is considerable, both in volume and in intervening transformations. Wackermann (1975) described recent alteration profiles from eastern Senegal, which could constitute a present-day model for such processes. It is to be noted that if the paleoalteration could be dated as 200–210 Ma, the uraniferous mineralisation attributed to hydrothermal fluids and often invading the paleoalterite is dated as 160–170 Ma. Here again, diagenesis could have modified certain characters acquired in the surficial environment.

2.8.2 Precambrian Paleoalterites

2.8.2.1 Limits of actualism
Meteoric transformations which could have intervened during the Precambrian require particular attention. It is rather difficult to compare Triassic or Paleozoic paleosols with present-day models; it becomes even *riskier* when comparisons are made with Precambrian alterites. The composition of the atmosphere and the selectiveness of biological activity indicate in no uncertain terms that the environment of that epoch was very different from that of today. Apart from the ancient nature of the rocks, the probability of **diagenetic, hydrothermal** and even **metamorphic readjustment** is high.

Some possibilities can exist nevertheless, as discussed in particular by Retallack, Grandstaff and Kimberley (1984) and by Retallack (1986). In the last few years much work has been carried out in this domain and a few authors have proposed fairly accurate interpretations. Bertrand-Sarfati and Moussine-Pouchkine (1983) described, for example, the fossil **calcretes** from Mali dated as 600–700 Ma; Kalliokoski (1975) also interpreted the altered layers from the Archaean of Michigan as calcrete, while some others prefer a hydrothermal explanation (Lewan, 1977). Gay and Grandstaff (1980) identified paleoalterites dated to 2400 Ma from Ontario and compared them to present-day **gleys** and **podzols**; Ross and Chiarenzelli (1985) identified silcretes in northern Canada dated approximately 1700 Ma. Schau and Henderson (1983) studied the Archaean formations in Canada and on the basis of prolific geochemical data concluded a relative permanence of the alteration phenomenon, which corresponds rather to *readjustment of the minerals of the rock among themselves* wherein the atmosphere is not of fundamental importance.

2.8.2.2 Weathering profiles in the Precambrian of Saskatchewan (Canada)
Athabasca basin, situated in north-western Saskatchewan state, has been well studied since it contains uranium beds of economic importance. An alterite aged about 1700 Ma has been recognised beneath the Athabasca sedimentary formations (McDonald, 1980). Two types of profiles in this alterite have been schematised in Fig. 56. It appears that the observed paragenesis occurred in several periods (Pagel, 1983; Halter et al., 1985): a regional **retrometamorphism** was followed by surficial alteration, and then by **burial diagenesis** and **hydrothermal activity**, resulting in the mineralisation.

Identification of meteoric paleoalterites depends on the following criteria:

— The subjacent sequence is clearly discordant; there was an emergence with the formation of a paleosurface.

— Alterite does not exist throughout; erosion must have been active on certain points of the paleosurface.

— Diverse characteristics are typical of evolutions with a ferrallitic tendency; there is a relative enrichment in Fe_2O_3, Al_2O_3 in the profiles from the bottom upwards, and clays of type 2/1 transform into type 1/1: sometimes pisolitic structures appear.

— The probable position of the equator in this epoch could have been 20–25°N latitude, which conforms well with the above proposed evolution.

All the characters of the profile may not be due to meteoric alterations, however; diagenesis has certainly been manifested in some minerals:

EXAMPLES OF PALEOALTERITES AND PALEOSOLS

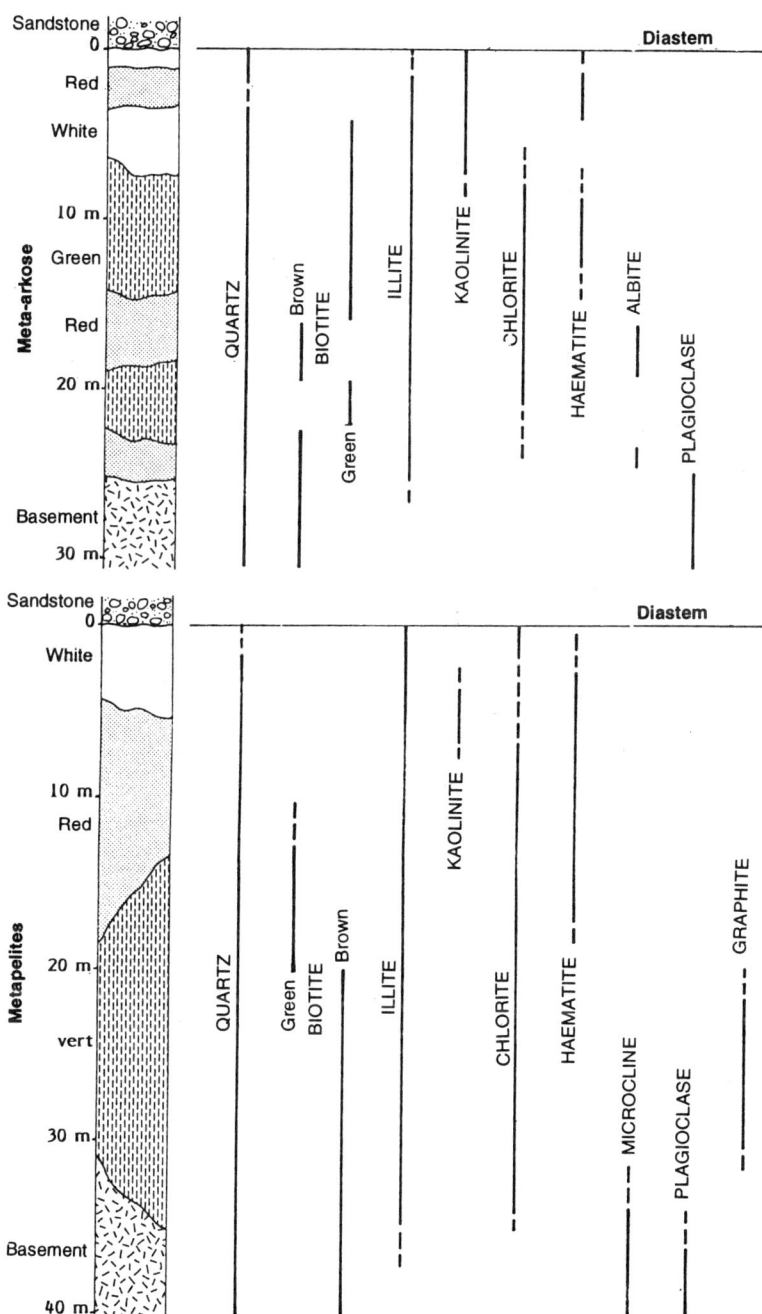

Fig. 56. *Facies and mineralogy of a Precambrian paleoalterite (after McDonald, 1980). Thickness of the alterite varies with the parent rock but the mineralogy remains somewhat similar. Precambrian of Saskatchewan.*

— Absence of smectite in the profile and good crystallinity of illite indicate an advanced diagenetic evolution.

— Iron oxide occurs in the form of **haematite**, a mineral with greater possibility for being a product of recrystallisation of **goethite** or of **limonite**.

— A study of fluid inclusions in minerals associated with silicification indicated the **temperature of formation** to be from 150 to 250°C; their salinity as well as the oxygen isotopic composition prove that these minerals were in equilibrium with diagenetic waters.

2.8.3 Diagnosis and Principles of Interpretation

Varied and changing characters of paleoalterites do not permit a precise definition; their investigation is difficult indeed and certain rules must be strictly respected:

— It must never be postulated that a paleoalterite is a product only of *surficial weathering*; *diagenetic transformation* must also be kept in mind by the investigator.

— Most of the recognisable paleoalterites have undergone an alteration with *ferrallitic tendency*, with destabilisation of type 2/1 phyllosilicates in favour of type 1/1 clays.

— When these appear on endogenous rocks, the progressive *destruction of primary minerals* helps in understanding them.

— Interpretations are based on paleopedological elements, even when these are very fragmentary; the **horizons of accumulation** are especially useful.

— In Precambrian paleoalterites *geochemistry* and *mineralogy* are the two fundamental tools of investigation.

— A study of the paleogeography envisaging the **source** zone and the zone of **sedimentation** could be the best method of evaluation, though it seems difficult to achieve.

3. PALEOALTERITES AND PALEOSOLS IN THE TECTONO-SEDIMENTARY CONTEXT

3.1 Factors Controlling Formation and Preservation of Paleoalterites

A concurrence of favourable circumstances is necessary for the formation of alterites and soils and for their fossilisation. The factors responsible for these conditions are variable:
— **tectonic** *context*, primordial factor;
— *topographic location*, which depends on a combination of tectonics, climate and geological substrata;
— *geochemical context*, related to climate and geological substrata;
— *dynamics of* **sedimentation**, a complex factor depending on the preceding ones.

3.1.1 Tectonic Context

3.1.1.1 Paleoalterites and global tectonics
The phenomena which model the lithosphere act at an infinitely larger scale than those which result in the appearance of alterites. The formation of alterites is nevertheless dependent on global phenomena. Analysis of cratonic stratigraphic sequences by Sloss and Speed (1974) helped to determine, in terms of probability, the tectonic disposition of the paleosols. These authors based their analysis on the deformation of cratons, which are never completely stabilised. They thus defined three modes of deformation which are connected with the events, especially on the fringe of the craton.

Cratons with emergent mode of deformation. Tend to rise progressively with respect to the sea level; erosion is regular and prominent discordances are excellently recorded, though there is little scope for the alterites to fossilise. These cratons are bordered by inactive margins where submarine sedimentation generally occurs.

Cratons with submergent mode of deformation. Progressively depressed and underwent major transgressive tendency. The calm environment is favourable for fossilisation of rare alterites which may develop. These cratons are often bordered by a zone of collision resulting in the development of considerable relief; erosion results in an increased rate of sedimentation which exceeds the receiving capacity of the subsiding border of cratons, which is gradually invaded by vast detrital spreading. The **Old** and the **New Red Sandstones** are examples rich in paleosols.

Cratons with fluctuating mode of deformation. If the general trend is regressive the elevation of cratons is spasmodic and tectonic faults simultaneously create a zone of considerable erosion and sediment traps. The quantity of sediments is enough to keep the continental traps always filled; consequently the existence of vast zones of continental sedimentation ensues; paleosols may appear and fossilise but only with local extension. These cratons are sometimes bordered by island arcs limiting the marginal basins, where the rate of supply of continental detrital material is so rapid as to cause an emergence and also the appearance of paleoalterites.

3.1.1.2 Paleoalterites and sedimentary basins

A global analysis of sedimentary basins was proposed by Kingston, Dishroon and Williams (1983). Among the parameters that the authors took into consideration, **sedimentary cycles** is an important one, with each cycle registering a tectonic event that could have left imprints during the evolution of the basin. A normal cycle of sedimentation commences above a **discordant surface** by *continental deposition,* gradually followed by *marine depositions* and deepening of the basin which in general induces marine transgression. The cycle is terminated by regression and return to continental facies. Fig. 57 schematises the characteristic deposition of the alterites which may fossilise under the influence of such a cycle. Such a scheme certainly constitutes the starting point for studies but several factors intervene and modify it:

— The unconformity at the top of the cycle may be due to considerable erosion which eliminated the upper parts of the deposits.

— The filling of a basin is generally accomplished in many cycles. Faulting causes the formation of horsts and grabens.

— The intensity of **subsidence** is in concurrence with the rate of **sedimentation**, which is related to the deformations in feeder zones.

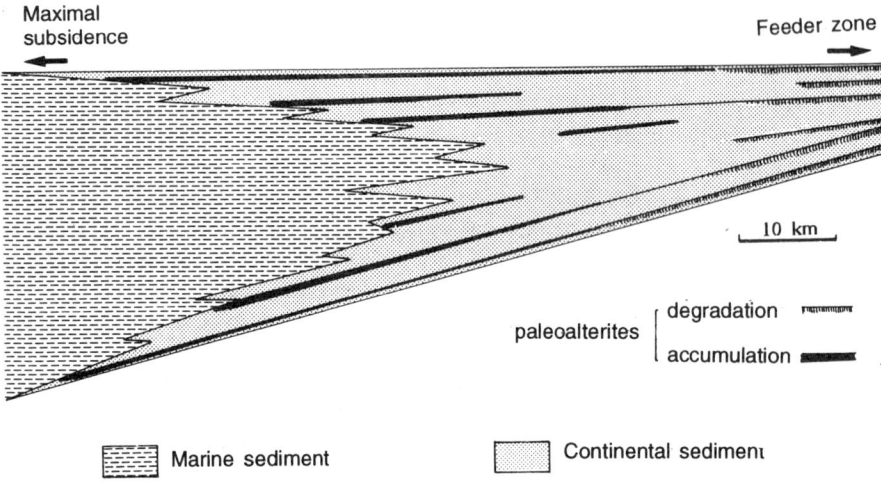

Fig. 57. *Theoretical sedimentary cycle in the margin of the basin. The vertical scale is highly exaggerated. The cycle commences and ends with a generalised paleoalterite; development of this paleoalterite is very limited and very irregular during the cylce. The paleoalterites are generally characterised by degradation upstream while accumulations prevail downstream.*

3.1.1.3 Paleoalterites and orogeny

Metallogenic analysis of **bauxites** has led to assigning them a definite tectonic position (Nicolas and Bildgen, 1979; Bardossy, 1981). Karstic bauxites of Europe, for example, are situated in orogenic belts. Most of them are correlated with mio-geosynclinal zones and, to a lesser extent, with eu-geosynclinal zones. It is difficult to date these alterites precisely as they are placed in periods of hiatus, which are often long. It is possible to relate them to stages of tectonic evolution, however; for the most part they were formed during *relatively calm phases which precede the main orogenic phases*. Based on this logic, Laville (1981) has stated that in the ensemble of the Pyrenees the bauxites in the inner zones, near the orogenic axis, are intercalated with the Lower Cretaceous sequences, while those in the outer zones are intercalated with the Upper Cretaceous sequences.

All correlations between paleoalterites and **tectonics** may seem schematic. In fact, it is apparent that the emergence and the continental sequences at the place where paleoalterites are localised, do not appear by chance in the geotectonic cycle. However, this statistical approach must not monopolise the attention of the geologist. Paleoalterites which have fossilised at a place not expected *a priori*, for example in the midst of a marine sequence, completely overthrow preconceived ideas, forcing the models to be questioned. This is uncontestably very valuable information for the **reconstruction of paleogeography**.

3.1.2 Climatic Context

Climatic zonation of soils on the surface of the earth has long been acknowledged. Pédro (1968, 1979) associated a *zonation of geochemical processes* with it (Table 3). Based on climatic zones the author distinguished two main processes: **acidolysis** wherein the intensity of weathering remains minor compared to **hydrolysis**, which is very active when the climate becomes warm, and especially humid. Table 3 is a tempting reference to geologists in search of elements for paleogeographic interpretations, but a few things about paleoalterites must be kept in mind:

Table 3. *Influence of large* **climatic zones** *on the pedological and geochemical processes (after Pédro, 1968, 1979). The occupied surface area is expressed as a percentage of recently emerged land.*

	Climate	Pedogeochemical evolution	Surface occupied	Secondary minerals	Soil types
A	Polar	Frozen zones	10%	None	
C	Northern	Podzolisation	16%	Solubilisation (three-layer clays) H	Podzols / Podzolic soils
I					
D					
O					
L	Temperate	Aluminisation	12.5%	(three-layer clays) Al	Podzolic ochreous soils / Acid brown soils
Y					
S					
I					
S					
H	Mediterranean	Bisiallitisation s.s	26.5%	(three-layer clays) Ca	Fersiallitic soils / Tropical brown soils / Vertisols
Y					
D					
R	Dry Tropical				
O	Humid Tropical	Monosiallitisation	18%	Kaolinite	Ferrallitic soils
L		Allitisation	13%	Gibbsite	Ferrallites
Y	Equatorial				
S					
I					
S	Absolute Desert	?	4%	None	

— These diverse types of alteration are modulated by the *degree of evolution* (Pédro, Delmas and Seddoh, 1975); for a paleosol, the degree of evolution is a function of the *time* during which weathering was effective.

— Highly alternating warm and humid climates have a greater chance of leaving imprints in an ancient sequence compared to less aggressive climates; this can result in a *distortion of the information* which may have been collected recently.

The logic followed in Table 3 is rationalised by *local factors*, which may gain prime importance; for example, the topographic position or geochemistry of the parent rock, which will now be discussed.

3.1.3 Topographic Context

Duchaufour (1977, p. 423) showed that in a humid tropical climate the following suite of soils is encountered on basic rocks: **ferrallite** with gibbsite on plateaus (**allitisation**), **ferrallitic soil** with kaolinite on the slopes (**monosiallitisation**) and a vertisol with smectite in the lower regions (**bisiallitisation**). These facts definitely complicate the scheme of Table 3. When fossilised beneath the later sediments, paleosols and paleoalterites can be classified into two groups as functions of two distinct modes of evolution:

— Those which develop during a general **emergence**; these signify a **paleolandscape subjected to decay**. There is almost no external supply of sediments, which allows for the paleosurface to evolve for a long period, so as to acquire a topography with varied relief and to be covered by equally varied alterites. This is brought about by large **discordance**.

— Those which develop in a landscape **subjected to accretion**. This occurs in a continental environment but there is a *periodic supply of sediments* and paleosols develop between two major phases of supply, on *vast flat topographies at a regional scale*. This flatness and the topographic disposition with its nuances are expressed in details: classical lateral changes in the paleosol facies and thinning and thickening of horizons of accumulation can be observed in the field. All these give rise to a suite of soils (**toposequences**) in which each profile is developed under specific conditions, receiving from the upstream and releasing downstream. Sedimentological analysis in most cases shows the recommencement of erosion at the channel level, the vertical amplitude of which is several metres. This implies a certain form of relief. Continental sequences developed in such landscapes are fossilised in a *succession of discontinuous paleosols,* relatively less evolved and rarely ferrallitic.

3.1.4 Geochemical Context

When the **parent rock** has a medium composition, for example granodiorite or sandstone cemented by calcareous material, the weathering likewise shows no particular orientation. This is not the case if the parent rock has a composition which is not common; *peculiar chemical elements* singularly orient all the intervening transformations in the surficial environments. Hence the following facts should be remembered:

— A determining factor is the presence or absence of calcium in the soils in which oxidation of *sulfides* occurs; if present, the **pH** value rises and gypsum is formed, while if it is absent the pH becomes very acidic and **jarosite** appears (Sec. 2.4.4).

— There is every reason to believe that the arid environments which prevailed in the north-eastern part of France during the epoch of formation of the Vosges Sandstone

(Sec. 2.5.3) were favourable for the formation of **evaporites**, yet no evaporite formed because neither the feeder zones nor the alluvial plains had the necessary chemical constituents.

A few important geochemical principles can be derived with respect to such environments:

— Existence of a **threshold** above which some reactions are possible. For example, *the threshold of sulfide content* in reductive environments stops the appearance of **siderite** in favour of sulfides into which iron preferentially enters.

— Available chemical stocks which determine certain transformations. For example, calcareous crust appearing on a substratum rich in calcite is formed much more rapidly than on a substratum devoid of calcium, and such an evolution is dependent upon the quantity of external supply (streams, wind).

— Mobility of chemical elements. The more soluble elements can **migrate** and accumulate rapidly at the favourable sites. However, a very high solubility causes multiple **remobilisations** and the disappearance of minerals from the landscape, where boxworks remain the sole witness.

3.1.5 Dynamics of Sedimentation

3.1.5.1 Sedimento-pedogenetic cyclothem

The concept of sedimento-pedogenetic cyclothem was proposed by Freytet (1964). The ideal case constitutes a sequence of deposits, one to several metres in thickness with positive graded-bedding, the diastem at the top corresponding to an exundation. The latter is sufficiently long to allow for *development of soil*, which is then fossilised beneath later deposits (Fig. 58a). These facts complicate this model:

— Thickness of alteration paleoprofiles may be greater than the sequences of the deposits, in which case the *paleoalterites telescope into each other*, involving all the complications presupposed (Fig. 58b).

— *Discrete sediment supplies* may intervene during the course of pedogenesis; these supplies are progressively 'assimilated' in the pedological profile and no clear diastem appears at the top of the paleosol (Fig. 58c).

— A **deposit sequence** or cyclothem must not be assumed as a simple evolution of sediments along the vertical axis; it is a manifestation of an event, for example a flood, causing the deposition of a *sedimentary body* of a certain volume. In other words, the problem has to be envisaged at the scale of the sedimentary body, which has been attempted in the following section.

3.1.5.2 Preservation of paleoalterites in alluvial sequences

The Old Red Sandstones of Wales are fluviatile deposits in which a number of **calcretes** of pedological origin are intercalated. Allen (1964, 1974) conducted studies of sedimentation of the sequences and megasequences of this formation and envisaged a relationship between the formation of *calcretes* and diverse types of fluviatile networks. Watercourses by and large migrate laterally but vertical adjustments are also possible as a function of the changes in the base level. These displacements are related either to the intrinsic characteristics of the watercourses or to some external factor, such as **tectonics** or **climate**. Allen proposed models in which all the possibilities were considered. Six cases of maximum practical interest are presented in Fig. 59. This model explains the formation and fossilisation of calcretes depending on the determination of three parameters: the *properties of watercourses*, the *tectonics* and the *climate*. This model appears to be transposable and

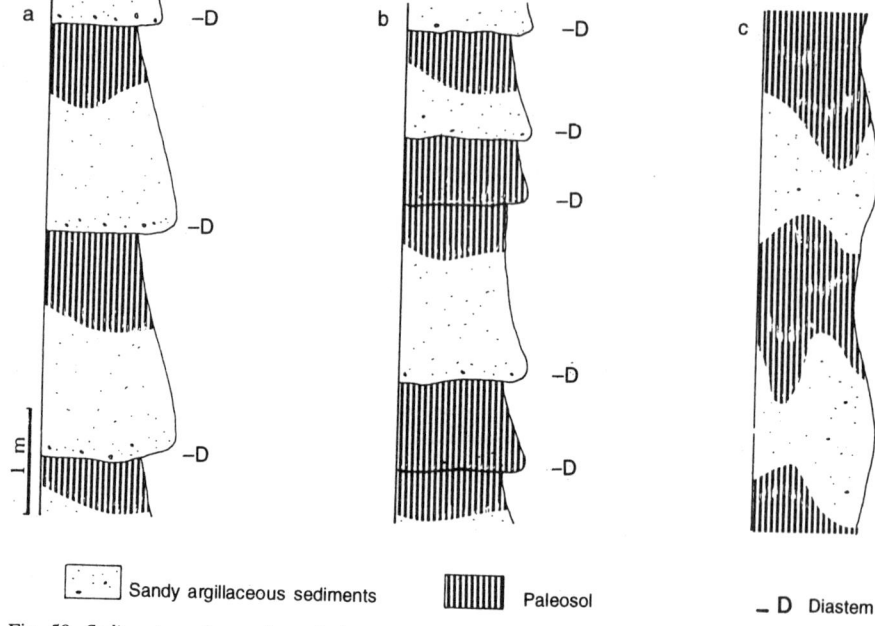

Fig. 58. *Sedimento-pedogenetic cyclothem (after Freytet, 1964).*

  a) Each diastem, at the top of a depositional sequence, is characterised by a paleosol.
  b) The very small thickness of the depositional sequence often results in the telescoping of paleoprofiles of alteration.
  c) A decrease in the rate of sedimentation is evident but not complete cessation; distinct diastem does not form.

amenable to generalisation, thus facilitating comprehension of the various types of paleosols appearing in alluvial plains.

One case in particular remains to be studied, that in which no paleosol is fossilised within an alluvial sedimentary formation. Such a case is not represented in Fig. 59 because it is very rare and can only happen in the context of at least one of the following:

— Very ancient alluvial sequence, for example Archaean, an epoch when *biological activity was limited.*

— *Alluvial sequence deposited in a desert environment* wherein soils did not form at all due to the climate.

— Alluvial sequence formed by a poorly organised fluvial network flowing over a plain periodically subjected to *sudden and devastating floods,* eliminating all traces of vegetation (this case is contrary to the meandering fluvial network whose migration is limited by the amplitude of the meanders and which is therefore contained within the zone of plain liable to flooding).

— Alluvial sequence deposited in a **rapidly subsiding basin,** in which there was no time for soils to develop between the sediments of two depositional phases.

— Alluvial sequence deposited in a **very slowly subsiding basin**; soils develop easily, but slight variations in the base level of the watercourses cause frequent reworking of the sediments during migration of the channels, thus precluding fossilisation of the soil (M. Durand, pers. comm.).

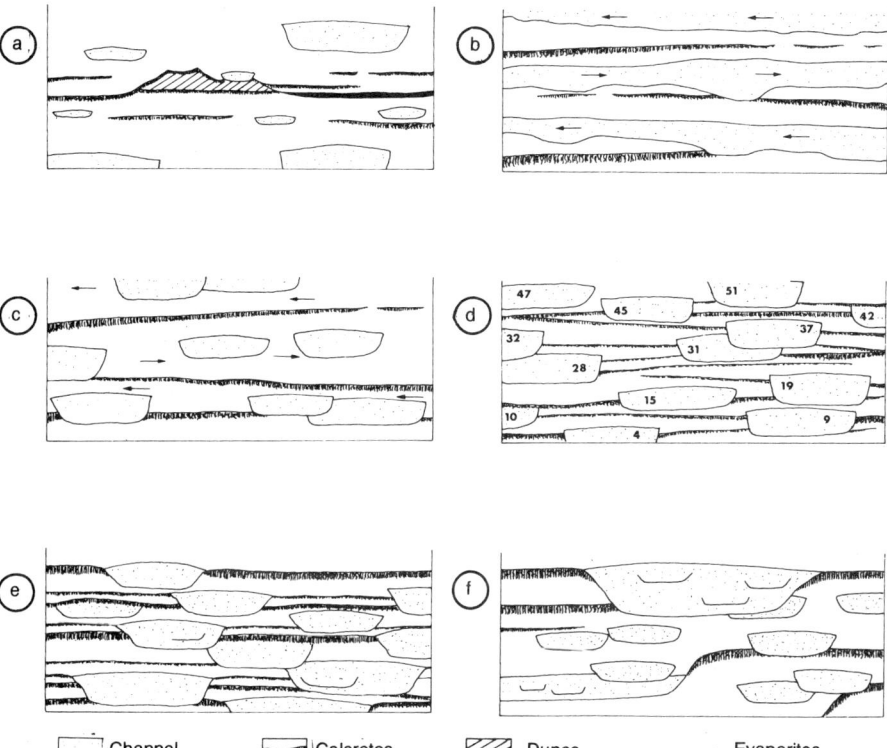

Fig. 59. *Relations between different types of watercourses and development of calcretes in an alluvial plane (after Allen, 1974). Sections are perpendicular to the general direction of water flow. Vertical scale enlarged 1000 times.*

a) Climatic fluctuation. During humid periods supply of sediments fills channels; progressive drying occurs at the origin of the paleosols, aeolian dunes and even evaporites (sebkha).

b) Continuous lateral migration of channels. Arrows indicate direction of migration.

c) Lateral migration of channels in 'steps'. Arrows indicate direction of migration.

In types b and c calcretes are developed at the lateral margins or the alluvial fan.

d) Lateral migration of channels in a disordered fashion. Numbers indicate order of channel filling. Development of calcrete appears to be anarchic.

e) Lateral migration combined with change in base level, the amplitude of which is less than the depth of the channel. Filling of certain channels is composite: calcretes are fairly uniform laterally.

f) Lateral migration combined with change in base level, the amplitude of which is greater than the depth of the channel. Filling of the channel is often composite; paleosols cover the interfluves up to the bank of the watercourses.

Kraus and Brown (1988) proposed a new approach to reconstructing ancient fluvial sequences using pedofacies analysis.

3.1.6 Formation, Development and Disappearance of Paleoalterites

The primary lesson of this section is the interdependence of most of the factors responsible for the formation of paleoalterites. The importance of **tectonics** should be emphasised: not only does it influence the *development* of alterites, it is also responsible for their *fossilisation*. Stable isotopes are giving useful information about paleosols (Goldstein, 1991) and their fossilisation.

The ideal condition for the formation of paleoalterites is as follows: an emerged landmass, tectonically stable, and a long period of warm and humid climate, thus creating considerable morphological and geochemical anomalies for them to be fossilised if there is a later submergent tectonic tendency, which is unfavourable to erosion.

The environmental situation in which paleoalterites are probably least expected, excluding oceanic basins, are the rising montane zones where **erosion** prevails over all other processes.

The *process of compensation* balances these extreme tendencies, however; the *intensity* of one factor may compensate the *weakness of another*.

— Under slightly aggressive climate, *duration* may compensate the weak intensity of alteration.

— Tectonic stability is a relative factor: the duration of formation of soils is short compared to the total time required for deposition of a sedimentary formation; even when the rate of sedimentation is high, interruptions in sedimentation are always possible.

— Superposition of many paleosols in the same sedimentary formation is common. These cannot be related to tectonic movements without considering the dynamics of watercourses in the alluvial plain.

— If a paleoalterite is subjected to erosion before burial, its preservation depends on *mechanical resistance*; this is particularly so for calcretes and silcretes.

3.2 Specific Location of Paleoalterites in Geological Formations

3.2.1 Paleoalterites Associated with Discordances

The presence of a meteoric paleoalterite below a discordant surface implies two fundamental conditions:

— the discordance might be related to an *emergence*;

— a *distinct erosion* intervenes before the deposition of sediments which fossilised the discordance.

Several examples are illustrated in this book, such as the paleoprofiles found on a Hercynian or Archaean basement (Sec. 2.8) or the calcrete emphasising an angular discordance between the **Chalk** and Cenozoic deposits in Champagne (Sec. 2.3.2.2). In general, a stratigraphic break is much larger than the time necessary for the formation of an alterite, so that only the most recent meteoric transformations are recorded in the paleoalterite.

In most cases an alterite is fossilised *below the discordance* but alterites do exist in the *first sedimentary layers overlying the discordance*. Examples are paleosols fossilised in the Wealden in the eastern part of the Paris Basin (Sec. 2.2.3). These facies, attributed to the Valanginian, rest on the Upper Jurassic marine limestones with an angular discordance. In an earlier study a chronological succession of events was proposed (Meyer, 1976), which is schematised in Fig. 60. A karst developed during the emergence which marks the Jurassic-Cretaceous transition in this region; it is a manifestation of a very efficient pedogenesis altering the Jurassic limestone. This paleokarst is not preserved everywhere; erosion has destroyed it at places together with the paleosols which covered it. This *erosion* gradually makes place for *alluvial deposition:* the first effect of the marine transgression is a rise in the base level; therefore watercourses progressively lose their capacity for erosion and *competence*. Fairly well-developed soils appear on the alluvial

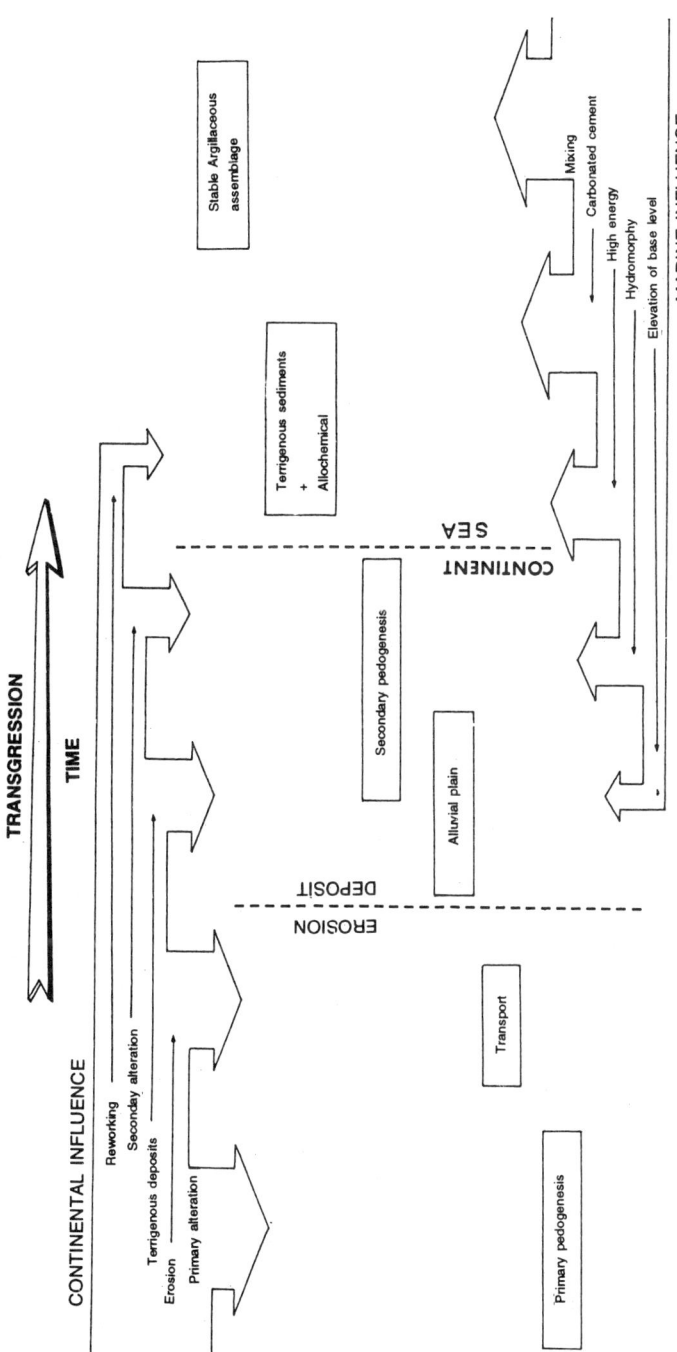

Fig. 60. *Model of Cretaceous transgression at the border of the Departments of Meuse and Haute-Marne. Schematisation of a single locality. Continental influence progressively disappears in favour of marine influence, which acquires more prominence. Two important breaks appeared in this progression: transition of the process of erosion into the process of deposition, and of continental environment into marine environment.*

plain because the weathering of sediments is vigorous (see Fig. 7a). A **hydromorphic tendency** develops, becoming primodial in the layers which precede the arrival of the sea. The first marine sediments (Hauterivian) indicate a high energy phase, which is represented by a *conglomerate* of Jurassic limestone pebbles with ferruginous elements derived from the Wealden paleosols and marine shells. The argillaceous matrix which encloses these elements gradually concedes to calcareous cements, reflecting changes in the chemistry of interstitial water. This marine influence becomes explicit with stabilisation of the argillaceous assemblage dominated by a well-crystallised **illite** (L. Barremian).

The fossilised paleosols in the first Cretaceous sediments of the region are a *precursor to marine* **transgression**, which measures the slow and progressive nature of the transgression. This is not a unique example: a fossilised paleoalterite below a discordant surface may present recurrences in the first sedimentary layers overlying the paleosurface. These recurrences are sometimes the only vestiges of paleoalteration.

3.2.2 Intercalated Paleoalterites in Sedimentary Sequences

Certain sequences, in particular molassic, may include a succession of almost continuous crusts or paleosols. A statistical estimation, made on the Oligo-Miocene **molasse** of Central Aquitaine (Meyer, 1981, p. 172), shows that on the same stratigraphic column a paleosol is observed statistically at 2.3 m interval and a well-evolved paleosol at a 4.3 m interval. The work of Allen (1974) on **Old Red Sanstones** converged to comparable results. The paleosols continue laterally for tens or hundreds of metres, occurring one after the other along the alluvial plain, and could serve as **indicators of paleoenvironment** and sometimes as stratigraphic markers.

Some intercalated paleosols in a sedimentary sequence may have a more specific significance, however, as shown by the following example. In the region of Rambervillers (Vosges) the Lettenkohle, which marks the transition between Muschelkalk and Keuper, can be described and interpreted in the following paleogeographic terms (Meyer, 1973, p. 55). The limestones deposited in the shallow waters of Muschelkalk are most often dolomitised, forcefully giving place to terrigenous clayey-silty material organised into a negative megasequence and terminated at the top by a paleosol containing *in-situ* plants (Fig. 61). The **sequence** indicates the progradation of a delta towards a very shallow sea during a calm tectonic period. Traces of vegetation are concealed by a coarse biodetrital limestone, rich in bony debris, which indicates beach deposits. This sequence evolved rapidly, limestone gradually giving way to dolomite and argillite to evaporites in Keuper. *The paleosol here marks the maximum regressive thrusts*; it also signifies a change in the environments. Muschelkalk sea where the Lower Lettenkohle was deposited, had a normal salinity. The influence of fresh water became perceptible during the build-up of a delta but transgression at the top of the paleosol resulted in a lagoonal environment with an increasingly evaporite tendency.

Here the paleosol had recorded an event in the basin. Such recordings continue sometimes for tens or hundreds of kilometres, similar to the example cited earlier, such as the Tertiary silcrete of Portugal and **violet zone** of Buntsandstein of Lorraine (Sec. 2.5.3).

3.2.3 Polyphased Paleoalterites

Some paleoalterites do not owe development of facies to a single phase of alteration, but to a *succession of evolutions* wherein the geological context is of prime importance. The following three examples indicate how numerous the possible cases might be.

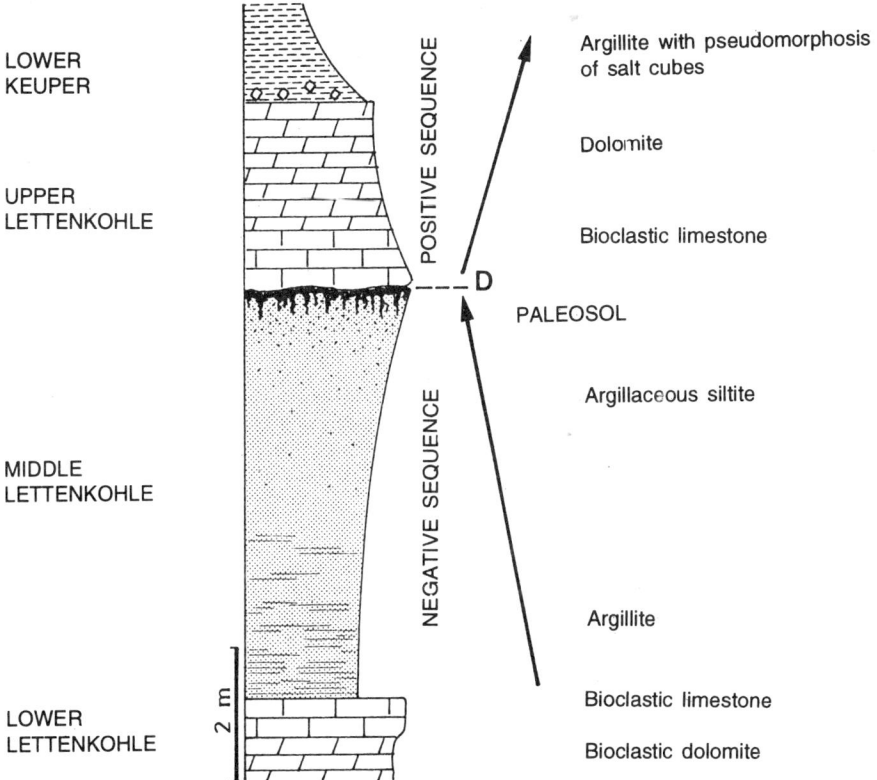

Fig. 61. *Section of Lettenkohle in the Middle Vosges. The sequential evolution shows that the main sedimentary diastem is characterised by a paleosol, which indicates a maximum regressive tendency in the environment.*

3.2.3.1 Bauxites on paleokarsts

In the cases described earlier (Secs. 2.6.2, 3.1) karstification of carbonates was accompanied by destabilisation of alumino-silicates, resulting in the accumulation of aluminium hydroxides. This took place in an oxidising environment. Hydroxides attained maximum concentration in the lower reaches of the bauxitised landscape, where a reduced lignitous layer had buried bauxite (Lecolle, 1967; Laville, 1972, 1981). Lignite was later oxidised. The sulfuric acid thus liberated deferrified the bauxite. The genesis of such beds is subjected to several constraints and in particular requires *two distinct phases of alteration.*

3.2.3.2 Outcrops of residual paleoalterites

Ferruginous paleocuirasses from West Africa (Sec. 2.6.1), though poorly dated, belong to the Tertiary. These constitute the core of numerous present-day landscapes but detailed studies (Leprun, 1979) show that they tend to undergo slow degradation. Their evolution is therefore polyphased; the great tectonic stability of the region in which they developed protected them from **erosion** and subjected them to *several types of climates.*

This example of a ferruginous duricrust is not unique. Comparable examples are found in the silcrete outcrops of Australia and Sahara. Generally old cratons with gentle relief and great tectonic stability favour the *perenniality of paleoalterites*, particularly when they are crusts (**silcrete, ferricrete, calcrete** and so forth).

3.2.3.3 Iron hats

In metallogeny, iron hats represent oxidised zones which often appear at the top of a metallic sulfide bed. Such hats are closer to the residual paleoalterite mentioned in the preceding section but may be indicative of deep-seated alterations, or several distinct periods of **oxidation**. The latter is due to the circulation of meteoric waters; the sulfides are destabilised, sulfuric acid is formed, and sulfates such as **jarosite** precipitated (see Fig. 19). Generally evolution does not stop here: as the sulfides disappear water becomes proportionately less acidic, while jarosite becomes unstable and transforms to **goethite**. The various metallic sulfates formed during oxidation are relatively soluble and facilitate migration of the metal towards the base, which may precipitate in large quantities above the water table. *These deposits of cementing material* may be sufficient in size for exploitation.

The development of iron hats is not systematic and depends on the regional context (Routhier, 1963, p. 242):

— In montane regions erosion may be so fast as to prevent their formation.

— In temperate climates the structures are most often similar to those of subjacent beds; the characteristic yellow or rust **colours** help identification in the field.

— In warm and humid climates iron hats are often confused with laterites or duricrusts because they may assume a slaggy or honeycombed appearance. Examination of the **boxworks, pseudomorphs** and **geochemical paragenesis** (Zeegers and Leprun, 1979) thus gains paramount importance.

3.3 Paleoalterites and Paleosols through Geological Time

Compared to most geological formations, paleoalterites and paleosols are smaller in thickness extension and are often fragile in character. More ancient alterites require greater attention to the study of *degradation of the facies*. Having set forth this principle, it appears necessary to examine the convergence of two periods for the development of paleoalterites on the surface of the earth: *the beginning of an oxidising atmosphere* and the beginning of *vascular plants*.

The atmosphere could have become progressively oxidising with the appearance of life (3500 Ma) essentially through photosynthesis during which oxygen is liberated (Retallack et al., 1984). The oldest **red beds** are known to be 2000 Ma, which clearly implies that the atmosphere was already oxidising in that epoch. Formation of paleoalterites older than 3000 Ma has been attributed to the action of meteoric water (Gay and Grandstaff, 1980; Retallack et al., 1984) but it is evident that our limited knowledge about the atmosphere of that epoch makes comparison with present-day models difficult. The rigorous **oxidation** encountered in present-day alterites cannot be assessed in very ancient profiles, which are known to occur below considerable discordances, which later became the sites for numerous diagenetic transformations (see Fig. 56). This does not imply a disregard for Archaean or Lower Proterozoic alterites; instead it is preferable to agree with Retallack et al. (1984) that considerable work remains to be done in this area and should be carried out with original methods since it is very promising. The aforesaid authors have even suggested that weathered profiles could have been more favourable for the *appearance of life* than the waters of the open sea.

During the Upper Proterozoic **meteoric** alteration seems to have been established from the actual physicochemical processes of recent times. Retallack (1981b) has opined that the large extension of very mature sequences which mark the infra-Cambrian and the Cambrian could be, at least in part, due to accentuation of continental weathering. One point remains obscure, however, namely the role of biological processes during this alteration. Vascular plants developed only in the *Silurian*; these differ markedly from those of today but it may well be that they were as significant then as they are today in the process of weathering.

The types of paleoalterites identified from the Precambrian to today appear to vary little; these are **laterites** (Boucot et al., 1974), or often siliceous or carbonated duricrusts (Bertrand-Sarfati and Moussine-Pouchkine, 1983). Almost all the types of paleoalterites described in this book appeared from the Paleozoic onwards, though their relative importance varies from period to period. It has already been mentioned that there were epochs especially favourable for the formation of bauxitic concentrations (Sec. 2.6.2); in this context it may also be mentioned that certain duricrusts *are favoured in specific periods*, for example **dolocretes** during the Triassic and **silcretes** during the Eocene.

In conclusion two important points should be noted:

— Alteration phenomena during the Precambrian require in-depth study.

— The possibility of degradation of the characters acquired during meteoric alteration, through time or through burial of the sequences, should be assessed.

SPECIFIC POSITION OF ALTERITES

Below a discordance:
Very long period of evolution, often polyphased, absence of erosion.

First layers above discordance:
Slow elevation of the base of watercourses, limited duration of evolution.

Materialisation of regression:
Little evolved pedogenesis, eventual influence of sea-water.

In a sedimentary basin, at the scale of basin:
Prolonged gap in sedimentation, stratigraphic marker.

Multiple occurrence in a continental sequence:
Climate favouring development of soil, moderate subsidence, well-organised fluvial network.

4. LITHOGENIC ROLE OF CONTINENTAL ENVIRONMENT

Meteoric alteration contributes to the development of specific rocks, both in present-day surficial environments and in **paleoenvironments**.

4.1 Transport of Material and Geomorphological Relief

Millot (1977) while discussing the dynamics of the mantle of alteration through time, proposed that it progressively subsided in the landscape. The *mantle of alteration grows from the base and is demolished at the top,* giving rise to two opposite geomorphological consequences:

— Crusts which indurate certain surficial horizons and simultaneously reinforce the landscape are *more resistant to erosion* and may fossilise beneath more recent sediments.

— Alterites are soft with high porosity and are concurrent with flattening of the landscape; this flattening depends not only on *mechanical* **erosion**, but also on the pedogenesis which becomes a genuine process of chemical erosion.

These observations can be transposed to past epochs. Only the intensity of the phenomena cannot be decided *a priori*; it must be determined separately in each case as a function of the **paleoclimate** and the **paleomorphology**.

4.2 Geochemical Barriers

Perel'man (1967) described a geochemical barrier as a lithological boundary wherein *the conditions for migration of fluids change radically. A geochemical barrier may be related to various geological phenomena, may be permanent, or may by functional only during a given epoch.* This definition has been adopted for diagenetic transformations and could also be extended to surficial environments. Some particular cases merit delineation.

— *Biological barriers:* Freytet (1971) categorised the trapping of terrigenous sediments by vegetation in inundatable plains as a *biological filter,* allowing certain zones of the basin to receive only the *dissolved products,* which resulted in very pure **palustrine** or **lacustrine** limestones.

— *Mechanical barriers:* the horizons of accumulation often constitute mechanical barriers whereby the permeability of the sediments is reduced sharply. The first consequence is an augmentation of the thickness of this accumulation, as often observed in calcareous crusts; the second consequence may be the accumulation of *foreign elements* in a situation wherein the effect of the water table and diagenetic water slackens.

— *Chemical barriers:* several alterites or pedological horizons may acquire the properties which convert these horizons into chemical barriers. A flux of air or aerated water in the rocks develops *oxidising barriers.* Horizons rich in organic matter and especially accumulation of vegetal remains (**histic horizons**) constitute, on the contrary, *reducing* *barriers* particularly wherever sulfides have accumulated. An *alkaline barrier* could also appear at the contact of carbonated or sulfated crust as a sharp rise in **pH** value results in precipitation of iron or manganese oxides.

4.3 Concentration of Minerals and Maturity of Sedimentary Rocks

Two types of mineral concentrations may result due to weathering of the landscape.

— *Residual concentrations*, which form as a result of the explosion of certain minerals or chemical elements from the profile, leading to relative concentrations of others minerals or elements (for example Fe, Al or Ti).

— *Absolute concentrations,* which are produced by the accumulation of external material or chemical elements within the profile (for example Ca or Si).

These concentrations are accompanied in all cases by an exchange of material which has an influence on the maturity of the rocks. Unfortunately, no general relationship exists between the phenomenon of alteration and the maturity of sedimentary rocks. However, such things do not occur by chance, but according to a few identifiable principles:

— At the scale of the profile, *biological activity,* which is often the only index of a fugacious pedogenesis, is brought about by the *mixing of facies* and consequently with diminishing maturity; on the contrary, a *very prolonged weathering* may result in an *almost monomineralic accumulation* and therefore a perceptible amelioration in the maturity of the rock.

— At the scale of **toposequence** the profile of weathering *receives* detrital elements and dissolved products from **upstream** and releases certain minerals **downstream**. Finally the *equilibrium* established in the profile between assimilation and expulsion determines augmentation or diminution of the maturity of the ensemble.

In the case of sedimentary rocks formed in continental environments, therefore, maturity depends not only on sedimentary processes, but also on paleoalteration which could occur sequentially on toposequences and during successive reworkings.

4.4 Chemico-mineralogical Characteristics Acquired in Surficial Environments

During burial **diagenesis** the rocks constitute a relatively closed milieu. Chemical reactions occur between the minerals and solutions, which are of finite volume, the replenishment of fluids being slow. Under these conditions reactions cannot continue in the same direction for long (von Engelhardt, 1977, p. 115). In general, a relationship exists between the nature of the precipitated minerals, the chronology of their appearance, the nature of the surrounding formations and the extent of burial. In a rock of continental origin, neoformations, on the contrary, could have intervened at the *surface* before any compaction; the chronology of authigenic minerals is generally complex. There exists a definite relationship between the chemical reactions and the nature of the formation, but many parameters have disappeared and are thus overlooked by the observer: topographic location, concentration of water etc. *These surficial milieus are definitely very open*; the solutions can vary within very wide limits, their replenishment is facile, and the composition may not be modified by chemical reactions. These environments are the site of *considerable allochemical transformations*. Table 4 presents a review of the neoformation of minerals in continental environments, fossilised in most cases where burial was negligible, thereby precluding burial diagenesis for the formation of these minerals.

Not all the minerals included in Table 4 are specific to continental environments but their high potentiality has to be admitted. Füchtbauer (1974, p. 323) further adds that meteoric waters often contribute to the lowering of concentrations in surficial environments, which favours the formation of large crystals.

This chemico-mineralogical approach to surficial environments would not be complete without mentioning one last point. The **redox potential** of continental waters is most often greater than zero and these waters moreover have a **pH** slightly less than 7, which leads to two important consequences:

— The oxidising character of the environment and the bacterial life result in *rapid mineralisation of the organic matter*; continental formations are most often poor in, or even devoid of organic matter. Convergence of very specific conditions is necessary for the formation of coal sequences.

— *The oxidising and acidic nature of water may persist* in the sediment up to a depth of 500 m to 1000 m, which induces diagenetic transformations different from those caused by interstitial waters of marine origin.

4.5 Physical Characteristics Acquired in Surficial Environments

Colour: Some colours are relatively specific to continental environments; those which are susceptible to fossilisation include:

— *Rusty-brown*, most often due to goethite.

— *Red*, in general due to pulverulent haematite, which may recrystallise to mallow or *lavender-blue*, colours characteristic of paleosols.

— *Green*, characteristic in pedological horizons rich in **ferriferous illite** (Durand, 1975).

— *Alternating stratified horizons of dark grey, clear grey, green or brown* indicating small variations in the degree of oxidation and/or the percentage of organic matter in fine sediments. These are characteristic of deposits in inundatable plains, which sometimes are the essential constituents of sequences extending over hundreds of kilometres, for example the Oligocene paleosols of South Dakota (Retallack, 1983) or the Triassic of Arizona (Meyer, 1984).

Hardness-Porosity: An early lithification which sometimes affects the horizons of accumulation, increases their density and hardness and decreases their porosity. Such cases, analysed by Yaalon and Singer (1974), are significant: on a homogeneous chalk of average density 1.42 and porosity 46% development of a calcrete results in a facies which gradually becomes denser up to the lamellar crust of density 2.49 and porosity 6%.

In the preceding example lithification affects a particular horizon but weathering can transform a sediment over a considerable thickness. In desert environments in particular **water-table** fluctuations are very large. The circulation of **vadose** water becomes primordial, causing alteration, deposition of cements at the contact of grains, or entrains migration and neoformation of argillaceous minerals. On the whole, continental sediments subjected to alteration in the environment of deposition acquired two properties which are found in the rocks:

— **Early cementation**, sometimes discrete, but sufficient to impart *mechanical resistance* to the sediments and therefore to undergo weak compaction during burial.

Table 4. *Minerals with their most common habits as they appeared in continental paleoenvironments. This schematic characterisation only expresses the tendency and is not exhaustive*

Mineral	Common habit	Paleoenvironment
Calcite	Micrite	Calcrete
	Sparite	Saline environment
	Rhombohedral (50 μm)	
	Rice grains	
Ferriferous calcite	Micrite	Pre-evaporite
	Sparite	
	Rhombohedral (50 μm)	
Slight magnesium calcite	Micrite	Calcrete
Dolomite	Micrite	Palustrine
	Sparite	Dolocrete
	Rhombohedral (50 μm)	
Ferriferous dolomite	Micrite	Pre-evaporite
Siderite	Sparite	Marshes
	Vadose pisolites	Organic matter
	Rhombohedral (20 μm)	
Gypsum	Lenticular	Gypcrete
	Desert rose	Pre-evaporite
	Rosettes	Evaporite
Anhydrite	Cleavages parallel to faces	Pre-evaporite
Jarosite	Rhombohedral (10 μm)	Oxidation of sulfides
Alunite	Rhombohedral (10 μm)	Oxidation of sulfides
		Silcrete
Celestite	Elongated prisms	Gypsic horizon
Barite	Lamellar	Presence of sulfate ions
Kaolinite	Foliated (SEM)	Mainly ferrallitic soil
Halloysite	Tubes (SEM)	Fersiallitic soil
Illite	Foliated (SEM)	Ubiquitous
	In laths by weathering (SEM)	
Ferriferous illite	Foliated (SEM)	Pluvial lakes
Smectite	Honeycombed (SEM)	Logged
Palygorskite	Fibrous (SEM)	Pre-evaporite
Sepiolite	Fibrous (SEM)	Evaporite
Phillipsite	Needles	Soil, pH > 8.5
Analcime	Globular forms	Soil, pH > 9.0
Goethite	Red-brown powder	Alternating desiccation-hydration
Haematite	Red powder	Alternating desiccation-hydration, warm
Gibbsite	White powder, often fine tablets	Humid and warm
Opal	Isotropic	Silcrete
Opal-CT	Fibrous	Silcrete
Quartz	Fibrous-cryptocrystalline	Silcrete
	Microcrystalline	Pre-evaporite
	Macrocrystalline	Acid sulfate soil

— Rapid loss in porosity, and especially of **permeability,** related to their cementation, but more so to the **argillaceous neoformations** which intrude the pores of the rocks (Wilson and Pittman, 1977).

Discontinuities: In sedimentary sequences paleoalterations occur at the origin of discontinuities appearing at diverse scales.

— The phenomenon of desiccation-humidification in materials rich in clays results in the appearance of vertical fissures, as well as *planes of oblique friction* (**slickensides**) along which polysilicates are oriented.

— Horizons of accumulation (clays, carbonates, silica etc.) become *less permeable* and often *very competent.*

— On a very vast scale the phenomena of weathering often give rise to relatively **permeable** rocks on discordant surfaces, possibly exploited by mineralising fluids, and hence *metallic beds in spatial relationship with such* **discordances** are very important.

PALEOALTERITES AND LITHOGENESIS

Types of geochemical barriers:
Biological, mechanical, chemical (oxidising, reducing, alkaline, acidic)

Mineral concentrations
— Residual or relative
— Absolute accumulations

Paleosequences
The profile of weathering establishes an equilibrium between the elements coming from upstream which are incorporated and the elements which are released downstream.

Authigenesis
A great variety of neoformed minerals, often large in size.

Physical properties of Paleoalterites
— Early cementation
— Poor permeability due to authigenic clays

5. METHODS FOR THE STUDY OF PALEOALTERITES AND PALEOSOLS

5.1 Study of the Outcrop

The geologist who suspects the presence of a paleoalterite in the field should confirm his hypothesis only after thorough examination of the paleoalterite with reference to the type and the objective envisaged. At the outcrop some precautions have necessarily to be taken to select the profile. *Pollution* ensuing from recent pedological evolutions *must not be underestimated* and the vicinity of the profile must be well apprehended: topographic location and local geology, degree of evolution of present-day soil, circulation of groundwater etc.

It is absolutely necessary to choose a profile which is preferably as distant as possible from the present-day surface of the soil, better still, from the lower limit or recent surficial formations. Studies based on deep drilling obviously set the geologist free from such constraints and the investigation is more reliable, albeit some loss of information concerning large structures is inevitable.

5.2 Investigation of Paleoalterites

A weathered intercalated layer within a sedimentary sequence is revealed by the study of transformations observed in the overlying rocks, the alterites and especially the apparently unweathered subjacent rocks. Changes in the structures and texture, mineralogy and geochemistry must be considered jointly. The methods applied for the study of recent alterites and the results obtained constitute valuable models (Tardy, 1969; Souchier and Lelong, 1970; Wakatsuki, Furukawa and Kyuma, 1977). However, transposition of the models cannot be direct because the solutions and amorphous phases described in them no longer exist. A total evaluation can be made on the basis of overall chemical composition of the facies and variation in mineralogical composition. The differential vulnerability of the minerals to weather results in their progressive disappearance from the base towards the summit of the **profile**. A basic model is provided in Fig. 62 which must, however, be used only in the proper context:

— Weathering is not always maximum at the surface but rather in the **zones of water-table fluctuations**.

— In a paleoalterite degradation of minerals may also be caused by circulation of diagenetic fluids.

— An investigation of authigenic minerals correlated with weathering is also very informative.

The results obtained are visualised through various schematic representations. The examples presented below indicate vast possibilities in this domain.

— Mineralogical evolution is sustained up to the level of an argillaceous assemblage, as described in several sections of this book (Figs. 7, 13, 49).

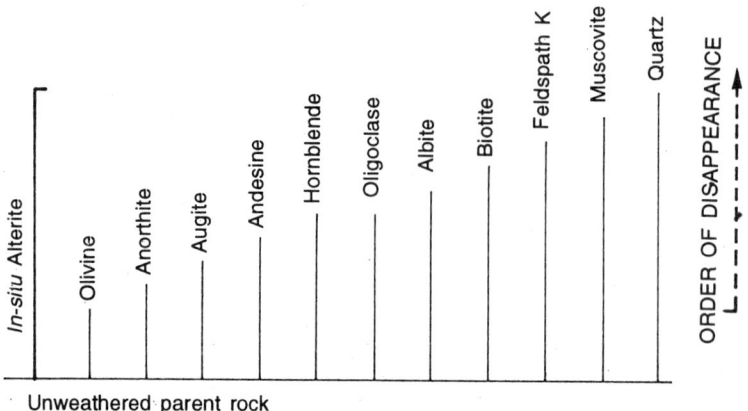

Fig. 62. *The common order of disappearance of principal primary minerals in an in-situ alterite (diagram based on data from Pettijohn, 1957, pp. 502–508).*

— Geochemical evolution is apparent in the variation of elementary aspects along a vertical section, for example as shown in Fig. 52.

The preceding appraisal can be improved if the elements are represented as a ratio of a known stable *invariant,* which allows the appraisal to be independent of the compositional changes in the parent rock, as well as variations in the comparative volume of the parent rock and the alterite. Unfortunately, a perfect invariant does not seem to exist in an open environment of weathering profiles; however, for convenience, Al, Fe, Ti are often considered to be retained as they are the least mobile elements in surficial conditions. Lelong and Souchier (1970) proposed the use of **quartz** as an invariant, considering a maximum of 15% chances of its disappearance. They made two calculations, one taking quartz as an absolute invariant and the other assuming it loses 15% of its mass. Gay and Grandstaff (1980) proposed the use of concentration ratios (CR) by taking Al as an invariant; they used the ratio as a function of the initial content of the element considered (% MO = mass of oxide of the element considered).

$$CR = \frac{\% \text{ MO alterite}/\% \text{ MO parent rock}}{\% \text{ Al}_2\text{O}_3 \text{ alterite}/\% \text{ Al}_2\text{O}_3 \text{ parent rock}}$$

$$= \frac{\% \text{ MO alterite}}{\% \text{ Al}_2\text{O}_3 \text{ alterite}} \times \frac{\% \text{ Al}_2\text{O}_3 \text{ parent rock}}{\% \text{ MO parent rock}}$$

An application of this representation is proposed in Fig. 63.

— Utilisation of more conceptual diagrams often improves the precision of interpretation. The classical triangular diagram (equilateral triangle representing three variables, the only constraint being $a + b + c =$ constant) could easily be adapted. The variants of a triangular diagram adapted for studying paleoalterites are shown in Fig. 64 (Chesworth, 1973; Chesworth, Dejou and Larroque, 1981).

Specific methods of investigation may furnish useful and decisive complementary information. A study of the composition of fluid inclusions trapped inside minerals enables assessment of the temperature of formation of these minerals; it also indicates the composition of the solutions from which the minerals crystallised. This method may be

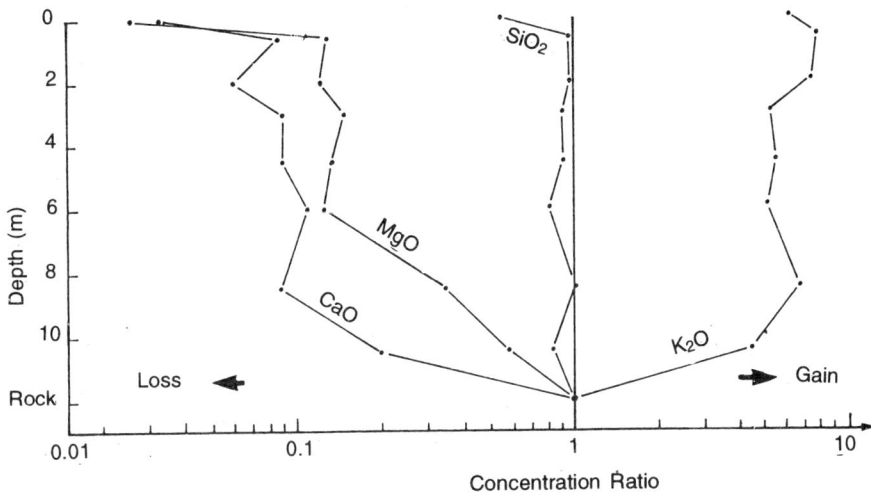

Fig. 63. *Evolution of some major chemical elements in a Precambrian alteration profile (after Gay and Grandstaff, 1980). Concentration ratio has been calculated taking Al as an invariant; the method of calculation is given in the text.*

applicable to neoformed minerals in an alterite, which are characterised by a temperature of formation less than 70°C. Relatively stable minerals, for example quartz, retain information during the course of diagenesis, but this is not totally true for minerals which recrystallise easily (carbonates) and in which the inclusions are in equilibrium with the last phase of recrystallisation. A study of the **stable isotopes** also enriches this knowledge. The most frequently used isotopic ratios are $^{18}O/^{16}O$ and $^{13}C/^{12}C$. It is nonetheless certain that the original values are susceptible to a complete change though the diagenetic history of the rock and hence interpretations are often more delicate. The stable isotopes of many elements can *a priori* be used to distinguish minerals and rocks of continental origin from those of marine origin (hydrogen, nitrogen, sulfur, silicon), which is a very practical problem. Analytical methods are cumbersome and a newly studied element can serve as a reference only after a very large number of determinations.

The results and the appraisal must be replaced in the ensemble of the mantle of alteration, which makes it necessary to revert to the field methods described earlier (Sec. 5.1): **cartographical control** is fundamental.

5.3 Investigation of Paleosols

The investigation of paleosols strives for maximum integration of the interpretable information as a function of existing models, in particular those which reflect criteria that enable comprehension of pedological mechanisms. It would appear logical to follow this procedure in the study of the paleosols:

— Analysis by the approved methods currently used by pedologists.

— Comparison of the results obtained with the characteristics of present-day soils for interpretations.

This procedure cannot be applied indiscriminately because there are systematic differences between soils and paleosols: the type of organic compound (C/N), exchange

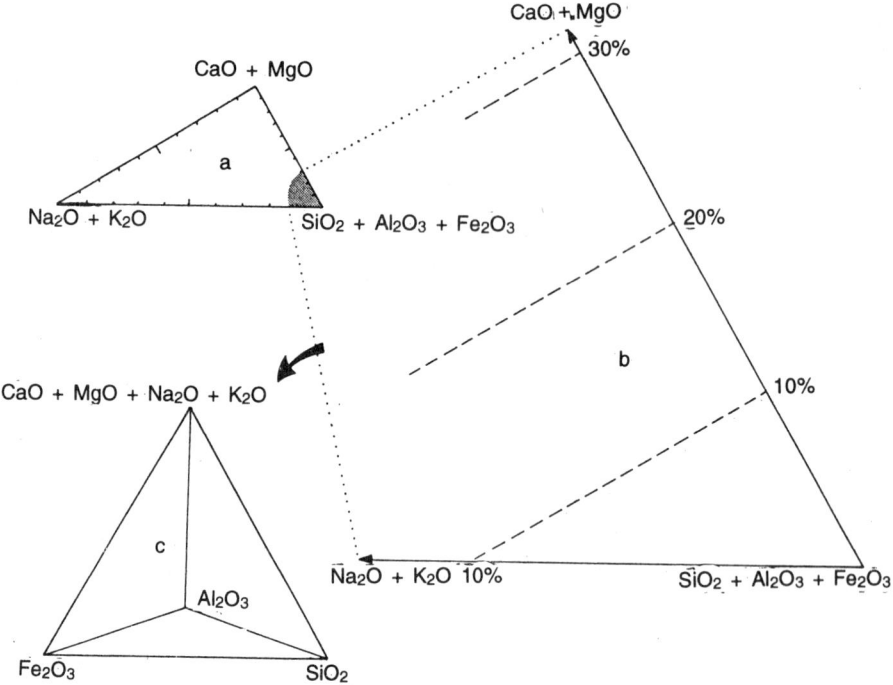

Fig. 64. *Examples of triangular and tetrahedral diagrams (after Chesworth, Dejou and Larroque, 1981).*

a) The equilateral triangle has been abandoned to bring into focus the $Na_2O + K_2O$ pole which actually represents a minimal content in the rock.

b) Enlargement of the zone which is most sensitive to leaching and alteration.

c) Tetrahedral diagram differentiating between the Si, Al and Fe; these three elements tend to accumulate in alterites.

capacity (T), rate of saturation (100 S/T), acidity and active limestones are examples of the important characters of a soil which *fossilises only moderately or not at all*. In several cases paleosols also have *truncated profiles,* surficial **horizons** having disappeared due to erosion before deposition of the overlying sediments. It is therefore necessary to remember the filter, constituting geological ageing, which transforms a soil into a paleosol; that is, an object with an individual identity requiring unique methods for investigation (Fig. 65).

5.3.1 Field Observations

When a paleosol is covered by a discordant deposit, it is easy to recognise the top. On the other hand, when it is enclosed within a formation, it may be difficult to delimit the *top* and the *base* of the paleosol; hence there is scope for planning the points of observation and extensive **sampling**. When there are several paleosols within the same sequence, *interference* is possible: many phases of pedogenesis could have affected the same sediment, thus increasing the complexity of interpretation of facies.

Nothing can replace field observations. The starting point is the search for relict textures of the host rock. In the case of sedimentary rocks development of a paleosol progres-

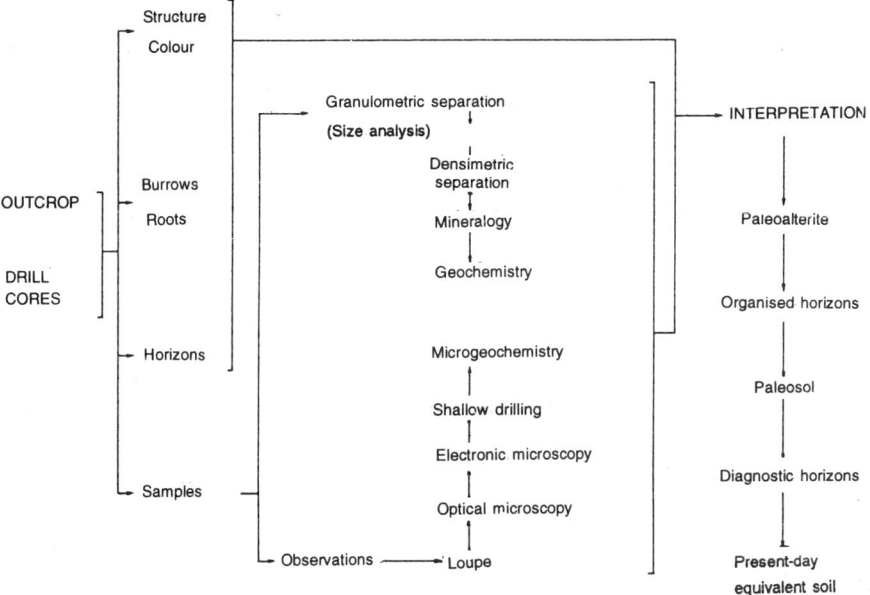

Fig. 65. *Organigram for the investigation of paleoalterite and paleosol. The method and technics are not limitative and should appear as a whole in the proposed general scheme.*

sively obliterates the depositional structures, which is quite revelatory. The **colours**, often varied, need to be determined with the help of a colour chart (for example *The Rock Colour Chart*), and all other specific structures (**prismatic or columnar structures, slickensides, nodules** and so forth) should likewise be noted.

Vestiges of biological activity such as **root** traces, **burrows**, **coprolites** must be looked for. However, it must not be forgotten that certain types of paleosols are completely devoid of all these characters.

5.3.2 Microscopic Observation in Polarised Light

5.3.2.1 Geological and pedological methods
The geological method envisages the textures, the types of a mineral and the criteria of reactions between minerals. It often enables visualisation of the role of interstitial fluids intervening in these reactions.

The pedological method, known as the **micromorphological** *method* or the study of **microstructures of soils**, does not neglect the minerals but is certainly more useful in the context of *textural relations*. This method was generalised and geologists began to take an interest in it after publication of the exhaustive work on this subject by Brewer (1964). The terminology of the author was accepted with reservations by pedologists, though by far the most utilised and even the only terminology appearing in geological literature. The terms defined in the glossary are a simplified version of this terminology, albeit several terms which figure in Brewer's repertoire have been omitted since they are barely identifiable or of no significance in geological sequences. One can also read the remarks given by Retallack and Wright (1990).

5.3.2.2 *Specificity of pedological microstructures, convergence*

All the microstructures given in Brewer's glossary are common in soils but for paleosols with a geological history, there may be a *convergence* with other phenomena. For example, **striotubules**, very common in soils, are sometimes observed in marine sediments; **argillans** develop by simple circulation of water containing argillaceous particles in a porous milieu (Brewer and Haldane, 1957); sedimentary orientation of clays imitate **plasmic separation;** mechanical diagenesis may also lead to plasmic separation of the **skelsepic** type. *Observation of pedological microstructures in an ancient geological formation does not per se prove the existence of a paleosol.*

5.3.3 Mineralogy and Geochemistry

The method of investigation used in petrology is naturally applicable and necessary for the study of paleoalterites and paleosols. A few precautions must be taken before analysis, however.

— **Sampling** at regular intervals of each defined facies; a sample of 100 g often suffices for ascertaining evolutions; complementary sampling at the limits of the facies helps to characterise the transitions.

—Whenever the facies is uncemented, separation of diverse granulometric phases presents a true picture of the facies and of evolutions; densimetric separation, though difficult in fine phases, completes the process for a precise approach (see Fig. 65).

— If **diagensesis** indurates the rocks to such an extent that any kind of separation is impossible, shallow drilling and electron-microscopic studies are the tools for chemical microanalysis.

5.4 Interpretations

5.4.1 Interpretation of Paleoalterites

• *Mineralogical and geochemical evolution*

Interpretation of paleoalterites primarily depends on the transformations of the **parent rock**. The selective degradation of silico-aluminates enables estimation of the intensity of weathering (see Fig. 62); clays which are the products of neoformation contain correlative information by virtue of their nature and abundance.

• *Hydric paleoregime*

The hydric paleoregime approach is most useful while searching for traces of emergence. *Phreatic paleoenvironments* are less specific because only the nature of the water which immerses them can differentiate the paleoalterites from the sediments deposited in a marine environment. A study of the stable **isotopes** in neoformed minerals may eventually be a good criterion. In the **vadose** paleoenvironment circulation of water could leave fossilisable traces such as cavities with geopetal filling, or elements presenting a polarised weathering or growth.

• *Paleomorphology*

Interpretation or paleotoposequences of alterites in the lithostratigraphic context leads to reconstruction of the landscape. In West Africa this approach has created doubt about the succession of erosion and alteration (Michel, 1978) and favoured an ancient relief telescoped by a single phase of Cenozoic alteration (Leprun, 1979).

5.4.2 Interpretation of Paleosols

5.4.2.1 Reconstruction of profiles

Recognition of different facies leads to the reconstruction of profiles showing superposition of several horizons, which should be compared with the type of soil. **Horizons** in present-day soils are designated by the alphabets A, B, C and G, subscripted with numbers or alphabets. A poorly differentiated horizon is denoted in parentheses, for example (B). The following definitions are retained:

— A_0: Surficial horizon made up almost completely of organic matter
— A_1: Surficial horizon containing a mixture of organic and mineral matter
— A_2: Eluvial horizon, poor in organic matter and often with clays and iron oxides leached out
— B: Illuvial horizon in which diverse products may accumulate:
  • humus: B_h
  • sesquioxides: B_s
  • clays: B_t
  • limestone: B_{ca}
— B_g: Illuvial horizon transformed into pseudogley, characterised by marmorisation of grey and ochreous patches
— G_r: Reduced gley, uniformly green
— G_o: Oxidised gley, with some rusty patches
— C: Parent rock, barely altered
— R: Parent rock

5.4.2.2 Inherent problems of paleosols

Surficial horizons (A_0, A_1) are generally not observed. They may have been eroded, transformed with the disappearance of organic matter, or 'assimilated' by the rising B horizon. However, no generalisation should be made. Paleosols with horizon A_1 are sometimes fossilised beneath volcanic ash (Yaalon, 1971) or *loess* (Schneebeli, 1976).

Paleosols by definition *are no longer functional*, which precludes any mechanism effective on present-day soils from having acted on them.

Degeneration of the characteristics acquired while the paleosol was functional is common; it varies depending on the nature of the paleosol and the conditions of fossilisation.

5.4.2.3 Utilisation of pedological classification in geology

Essentially descriptive classifications are those by FAO (UNESCO) and those by the United States Department of Agriculture, primarily based on the identification of horizons with well-defined characteristics termed **diagnostic horizons**. *Soil Taxonomy* (Soil Survey Staff, 1975) is a treasure-house of information in this domain. The use of diagnostic horizons is a seductive tool for attempting a classification of incomplete pedological paleoprofiles.

Genetic classifications, the French classification, and its variants proposed by Duchaufour (1977) are essentially ecological. These classifications are based on the following principle: the *environment* induces the *process*, which in turn causes the appearance of *characters* (Schroeder, 1973). Geological interest in the principle is obvious: starting from the characters, the geologist aspires to arrive at the processes and the paleoenvironment (this aspiration is questionable because of the relationship which is not strictly bijective). The concept of diagnostic horizons also enables attempting genetic classifications (Duchaufour, 1976, 1977, 1984).

Finally the following can be proposed:

— Identification of diagnostic horizons in the **paleoprofiles of alteration** through its characteristics.

— Utilising these diagnostic horizons as an interface, a direct attempt can be made at a descriptive classification and an indirect attempt for a genetic classification.

5.4.2.4 Utilisation of diagnostic horizons

The USDA classification defines diagnostic horizons with great precision; numerous criteria have been enlisted yet many are found to be totally missing during investigation of a fossil soil. Thus *diagnostic pseudohorizons* are often proposed in paleopedology. Even if total strictness in applying these characters to present-day soils is not possible, one should nevertheless adhere to them as closely as possible during investigation of paleosols.

For each horizon described below, the estimated frequency and probability of occurrence in ancient sequences are indicated in parentheses. Only those horizons which are of geological interest are discussed.

— *Surficial diagnostic horizons* (rare)

Mollic Horizon—A_1 (rare): Very dark horizon, containing more than 1% organic matter (thickness greater than 10 cm on hard rocks and more than 25 cm on soft rocks).

Histic Horizon—A_{01} (rare): horizon containing more than 20% organic matter; peaty, traces of hydromorphy.

Fundamental diagnostic horizons (common)

Argillic Horizon—B_t (common): presence of argillans of illuviation or stress cutans. Thickness at least 1/10th of the profile or 15 cm. Augmentation of clay compared to horizon A is difficult to establish since the profile is truncated.

Cambic Horizon—(B) (common): precursor to argillic horizon, shows an incomplete weathering of primary minerals, some traces of illuviation and beginning of leaching of limestone, if present.

Oxic Horizon—(B) (common): horizon at least 30 cm thick in which all the primary minerals (except quartz) have been weathered; rich in kaolinite, possibly in gibbsite.

Spodic Horizon—B_h and B_s (possible): horizon of organic matter and Al and Fe oxide accumulation.

Natric Horizon—(possible): variation of argillic horizon, with columnar structure and clays rich in exchangeable Na.

Secondary diagnostic horizon (some are very common)

Fragipan (possible): silty horizon, compressed and compact.

Plinthite (common): marmorised horizon with bright red spots; desiccating phases may harden the Al and Fe oxides. Paleosols which have continued in a phreatic milieu tend to become ochreous and hydroxides replace the oxides.

Albic Horizon (possible): eluvial, decolorised, develops typically at the top of a spodic or argillic horizon.

Calcic Horizon (common): horizon of carbonate accumulation (at least 5% carbonate greater compared to the subjacent horizon). Carbonate may occur as nodules or pulverulent spots.

Petrocalcic Horizon (common): indurated by carbonates. Hardness increases with respect to the calcic horizon and becomes greater than 3 on the Mohs scale of hardness.

Gypsic Horizon (common): rich in calcium sulfate (at least 5% more than the subjacent horizon).

Sulfuric Horizon (exists): contains jarosite or alunite.

Salic Horizon (possible, if preserved in the environment of fossilisation): horizon rich in soluble salts, NaCl.

Duripan (common): horizon indurated by silica minerals.

In sedimentary sequences the most easily identifiable horizons are those of **accumulation**; many belong to the class of secondary diagnostic horizons. Traces of **hydromorphy** have not been taken into consideration while defining diagnostic horizons; this may be advantageous since burial in a phreatic milieu affects the oxides of iron and manganese. Some traces of hydromorphy are related to the paleosol itself, however, and may be used in investigations.

5.4.2.5 Systematics of fossilised paleosols in ancient sequences

Tentative studies have been done in this domain. For example, Yaalon (1971, p. 34) tried to estimate the probability of their formation, in addition to the frequency of their observed pedological indexes. Freytet (1971) proposed a schematisation for the soils developed on carbonated mud.

The examples given in Table 5 constitute an almost complete review of diagnostic paleohorizons and corresponding paleosols. In the opinion of pedologists it is very simplistic, but may nevertheless orient geologists towards the possible types of paleosols. However, it should be noted that the types of paleosols are determined on the basis of diagnostic paleohorizons, which are the result of primary interpretations.

Table 5. *Diagnostic horizons* and equivalent soils which may be encountered in ancient sedimentary sequences. The thickness of the profile conforms to the observations made: *compaction* could be considerable. The association of some horizons in this table is similar to that presently encountered in the field.

Horizons	Equivalent soils	Profile thickness	Climate	Particulars
Cambic horizon -(B)-	Temperate brown soil Tropical eutrophic brown soils	< 1 m 1 m or +		Young soil, little evolved
Argillic horizon -B_t-	Leached brown soil Fersiallitic red soil	1 m 1 m 1 m	Temperate Tropical or Mediterranean	2/1 clays 2/1 > 1/1 clays
	Fersiallitic acid soil Tropical ferruginous soil	> 2 m	Tropical or Mediterranean	Very intense leaching 1/1 > 2/1 clays
Marmorised argillic Horizon -B_{tg}- Argillic horizon -B_t-	Fersiallitic acid soils with pseudogley	1 m	Mediterranean	Hydromorphy resulting from sealing of B_t

(Continued on next page)

Horizons	Equivalent soils	Profile thickness	Climate	Particulars
Cambic horizon -(B)- Calcic horizon -C_{ca}-	Little differentiated maroon soil	0.6 m	Mediterranean	
Argillic horizon -B_t- Petrocalcic horizon -C_{ca}- Petrocalcic horizon	Fersiallitic red soil with a Calcic horizon Soils with calcareous crust	1 m 0.2 to 3 m	Mediterranean Semi-arid to arid	CALCRETE (DOLOCRETE)
Calcic horizon or Gypsic horizon	Sierozems	1 m	Arid	
Petrogypsic horizon	Gypseous soil with a crust	1 to 2 m	Arid	GYPCRETE
Duripan		0.1 to 5 m	Arid tendency	SILCRETE
Oxic horizon -(B)-	Ferrallitic soil Ferrallite	2 m or + 2 m or +	Humid tropical Humid tropical to equatorial	1/1 clays Al and Fe oxides 1/1 clays
Oxic horizon -(B)-	Ferrisol with plinthite	1 m or +	Humid tropical to equatorial	
Plinthite	Ferrallitic soil with Petroplinthite	1 m or +	Humid tropical to equatorial	FERRICRETE (or ALCRETE)
Histic horizon -A_{01}-	Peat	2 m or +	Temperate	Permanent hydromorphy Considerable compaction
Albic horizon -A_2- Spodic horizon -B_h, B_s-	Podzolic soil Podzol	0.4 to 1 m 0.4 to 1 m	Boreal to temperate	
Mollic horizon -A_1-	Vertisol	1.5 m	Contrasting tropical or Mediterranean	
Argillite with swelling minerals	Pelosol	1 m	Contrasting temperate	
Sulfuric horizon	Saline soil with sulfate reduction	< 1 m		Alternating saline and fresh water
Natric horizon	Solonetz	> 1 m	Cold and dry	High pH
Horizon with aluminium phosphate	Andosol	1 m		On volcanic rock

The difficulties inherent in the nature of paleosols should be dealt with carefully: *all interpretations must be based on a combination of arguments*. As far as possible, these topics should be chosen from different scales (for example, *pedological microstructure* supplemented by the *succession of horizons in the field*) or, from a different aspect (for example, *geochemical differentiation* superposed on *differentiation in horizons*).

A necessary final remark: Development of present-day soils is more often a function of the lateral events along *toposequences*. Unless a large number of close outcrops are encountered or close drilling is carried out, both exceptional, it is difficult to ascertain the lateral variation in buried fossil formations. Hence pedological paleoprofiles can only be envisaged in vertical sections but this is not the only facet of reality. Cartographic data has also to be interpreted.

6. KNOWLEDGE ACQUIRED FROM PALEOALTERITES AND PALEOSOLS

6.1 Diagenetic Degradation of the Data Recorded in Paleoalterites

Like all other rocks, paleoalterites are obviously subjected to *diagenetic transformations*. To understand and to interpret a paleoalterite more precisely, it is necessary to know its diagenetic history. To comprehend the diagenetic history of a continental formation requires knowing how to correctly interpret the paleoalterites enclosed within it, obviously a theoretical impossibility. In practice, the geologist's approach can only be through successive approximations, taking into consideration the maximum data possible, which enables him to systematically acquire knowledge about the environment of deposition and the **conditions of diagenesis**.

An evolved soil may be approached through *paragenesis,* wherein the mineral and aqueous phases are in *equilibrium* with the environmental parameters, in particular pressure and temperature. As long as the conditions remain perceptibly unchanged, the mineral assemblages evolve no further. The soil becomes a paleosol, which remains without appreciable modification for long periods. The period may be as long as millions of years or even more. Ostensibly, fine carbonated microstructures or fragile clay, for example palygorskite, persist in the formations buried to shallow depths from the Paleozoic onwards. With *appreciable burial,* the equilibria contrarily are distinctly modified; the paleosol is gradually transformed and the *pedological features become progressively blurred.* Fig. 66 shows how the problem could be approached.

— The first transformation appears in the *scale of the crystals*. These transformations are mainly the dissolution and recrystallisation of relatively soluble minerals (sulfur, carbonates etc.).

— Thereafter degradation is effective at the *microscopic scale*, with the alteration of pedological microstructures, particularly plasma separations and cutans.

— Compaction and migration of fluids then transform the *microscopic aspect of the profile.*

— Further advancement of diagenesis, to the limit of metamorphism, more often affects only the total geochemical characteristics at the *scale of the paleosurface*.

— Outcropping after *burial* reintroduces pedological features in the profiles which are foreign to the environment of formation; these are true artifacts.

So the study must be adapted to the problem, which entails optimum observations. As a general rule, the deeper the burial of the paleosol the more necessary to study it 'from afar', even at the cost of a little precision.

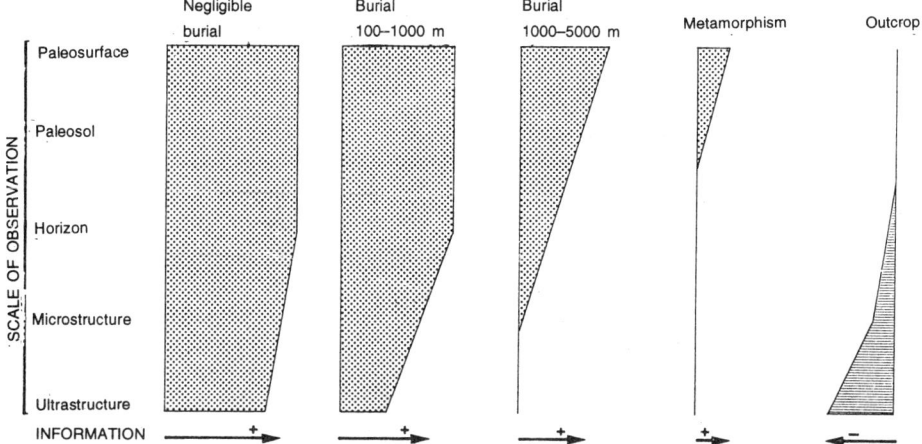

Fig. 66. *Information conveyed by paleoalterite and paleosol as a function of **burial**. This diagram visualises the phenomenon in a purely qualitative manner. Information diminishes progressively but disappears more rapildy with respect to small structures rather than larger ones. Outcropping detaches the system, introducing artifacts in it schematised as negative information.*

6.2 Knowledge of Paleoclimate

6.2.1 The Problems

Equilibrium of soil-climate-vegetation: **climax**
A climatic soil or *pedoclimax* is a soil sufficiently evolved to be in equilibrium with the general climate and natural vegetation. The **parent rock** thus plays only a secondary role (Duchaufour, 1977, p. 115).

Fossil soil may provide direct information about the paleoclimate if it attains this equilibrium while buried. Unfortunately, this does not always happen; besides it is also difficult to prove. The geologist is therefore obliged to interpret a system with several independent variables.

Sensitivity of paleosol to climatic variations
There is a **limit** (evolutionary threshold) below which a paleosol does not contain traces of climatic changes. The remarks of Barriere (1971) pertaining to Quaternary paleosols are transposable to ancient sequences:

— *Climatic thrusts* very often cause an acceleration or deceleration in pedogenesis, of which only overall results are available.

— Factors of pedogenesis change sufficiently only in the *marginal regions of larger* **climatic zones**, for such modifications to be recorded in the paleosols.

6.2.2 The Possibilities

It is evident that paleontological remains are better tools for precise interpretations compared to paleopedology. But the latter may furnish *limits or thresholds*, which even the paleontologist is obliged to respect. A significant example is the Aquitaine Miocene.

Ginsburg (1974) has noted that in the Sansan bed (Gers) four species of rhinocerotides cohabitated, whereas no such cohabitation occurs in nature today. To explain this observation he proposed the presence, locally at least, of a vegetal biomass 'superior to the great present-day equatorial forests'. Paleopedology refutes this hypothesis. The paleosols recognised in the Aquitaine Miocene provide no basis for envisioning a vegetation even at the scale of the equatorial forests (Meyer, 1981). Studies done by Retallack (1991) go the same way.

Besides determination of the thresholds, there are azoic continental formations and here paleopedology becomes one of the rare methods for determination of paleoclimates. *Climatic zonation* (see Table 3) and determination of *diagnostic horizons* (see Table 5) remain the basis of interpretation. The schematic model given in Fig. 67 is a complement since it enables in particular estimation of the **degree of evolution of the paleosol**.

Lithified **horizons of accumulation** are valuable auxiliaries in paleoclimatic research. It has already been mentioned that they have a very high probability for fossilisation. Fig. 68 shows the distribution of surficial crusts as a function of **climatic drainage**, i.e. *essentially of pluviometry*.

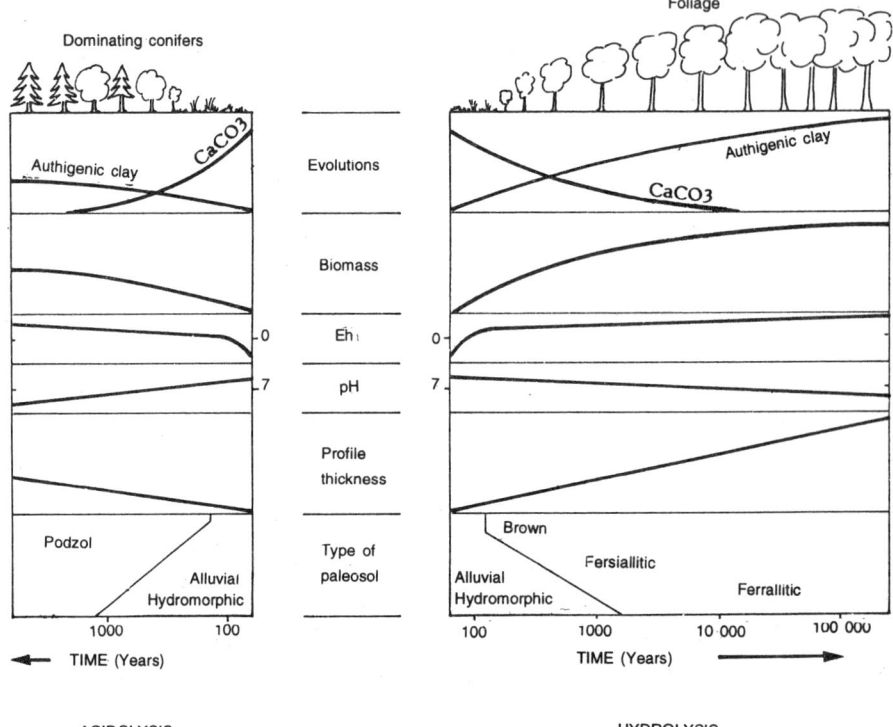

Fig. 67. *Theoretical model illustrating the two main processes of pedogenesis (inspired by G. Retallack, 1984). The soils supposedly developed on transported alluvial material of molassic composition. The evolution envisaged on the left (acidolysis) takes place in temperate humid to cold climate. Evolution shown on the right (hydrolysis) in humid and warm climate. These two diagrams illustrate that the features recorded in paleosols change as the duration of evolution increases.*

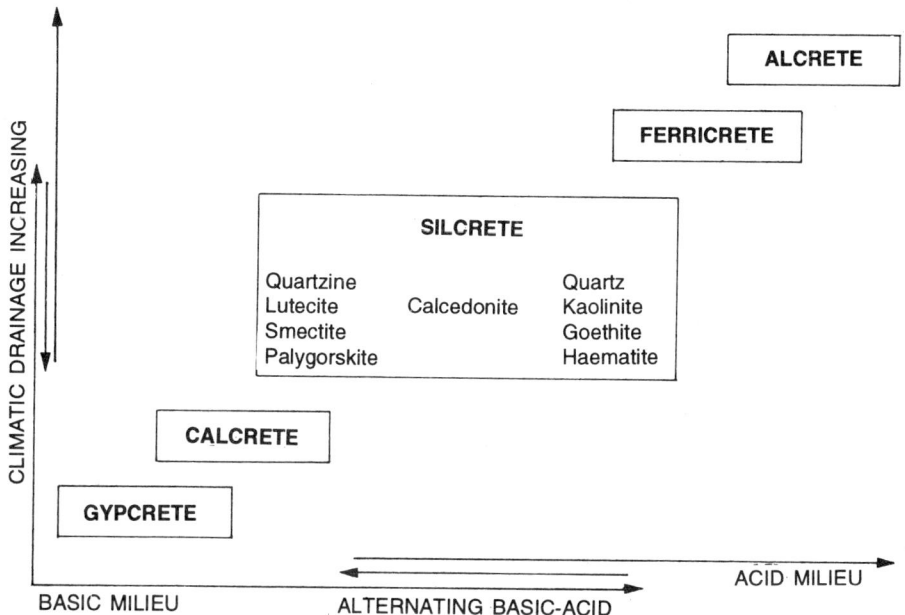

Fig. 68. *Nature of crusts or horizons of accumulation as a function of **climatic drainage**. Climatic drainage is defined as the difference between precipitation and potential evaporation-transpiration. Siliceous crust, which forms in various environments, may be distinguished on the basis of the mineral content.*

6.3 Knowledge of Duration

6.3.1 Paleosols as a Unit of Time

If the **time** necessary for the development of a particular soil type is known, the corresponding paleosol in an ancient sequence indicates the *minimum duration of the break* in sedimentation that allowed its development. The necessary time sufficient for the differentiation of a soil is not always easy to estimate, however. Much of the literature on pedology deals with estimation of the **age of present-day soils** but rarely precisely states whether the soils are still in the process of evolution, have attained equilibrium (**climax**) or have surpassed it; the geologist must approach age estimations found in pedological literature with scepticism. Table 6 presents the average ages as furnished by pedologists; the numbers represent the duration of formation of alterites, soils or even particular horizons. Although the significance of the Table has been accepted by the majority of specialists, a few remarks are imperative:

— the values are often not accurate;
— the time of formation may vary considerably for the same type of soil or horizon (for example, the calcretes).

Based on the values given in Table 6, identified and recognised paleosols can provide information on the duration of formation. These represent the minimum durations since soils after attaining equilibrium evolve no further.

Table 6. *Average time required for the formation of certain soils and pedological horizons. Three processes are apparent: Very rapid, such as the traces of hydromorphy, often associated with root layers; short cycles of monocyclic soil wherein soil-vegetation equilibrium is attained in 1000–10,000 years; long cycles, often warm climate, wherein evolution continues for tens, even hundreds of thousands of years*

Pedological phenomena	Duration of formation (years)	References
1. *Traces of hydromorphy*		
Redox variations	< 1	Veneman et al. (1976)
Gley	< 100	Duchaufour (pers. comm.)
2. *Short cycles*		
A_1 Humus-bearing	600–1500	Duchaufour (1977)
Decarbonation (polder, Holland)	250	Buol et al. (1973)
Decarbonation (dunes, England)	300	Buol et al. (1973)
(B) Cambic on carbonate rock (temperate climate)	3000–5000	Duchaufour (1977)
(B) Cambic on granite (temperate climate)	10,000	Duchaufour (1977)
Leached brown soil on loess (oceanic climate)	6000–8000	Duchaufour (1977)
Podzol on andesite	400–2000	Yaalon (1971)
Podzol of degradation (sandstone, Vosges)	1000–2000	Duchaufour (1977)
Podzol of degradation (oceanic climate)	2000–3000	Duchaufour (1977)
Podzol below leafy forest	4500–5000	Guillet (1972)
Vertisol	10,000	Buol et al. (1973)
Chernozem	2000–6000	Yaalon (1971)
Sierozem (Central Asia)	4000	Rozanov (1951)
1 m Calcrete (poor drainage)	1000	Netterberg (1978)
3. *Long Cycles*		
1 m Calcrete (good drainage)	1,000,000	Netterberg (1978)
Bauxitic ferrallite (Hawaii)	10,000	Valeton (1972)
1 m Ferruginous horizon (Senegal)	750,000	Nahon and Lappartient (1977)
Complex ferruginous duricrust (Senegal)	6,000,000	Nahon and Lappartient (1977)
1 m Alterite (India, monsoon)	55,000	Troy (1979)
1 m Ferrallitic soil	75,000	Buol et al. (1973)
1 m Laterite	60,000–2,000,000	Goudie (1973)
Silcrete	100,000–1,000,000	Summerfield (1983)

Three orders of duration which should be distinguished:

— Traces of hydromorphy appearing in a *few tens or hundreds of years*, which are abundant.

— Paleoprofiles which required *1000 to 10,000 years* to differentiate and which are the true paleosols are common in many sedimentary sequences.

—Highly evolved paleosols which took 10,000 to 1,000,000 years to develop emphasise, in general, large **paleosurfaces** of discordance.

The *power of resolution* of paleosols as a means for measuring the duration merits discussion. A chronozone, at best, is a duration of the order of 1 Ma. The subzone and horizon measure in biostratigraphy to a precision of hundreds of thousands of years. The durations in Table 6 indicate that the paleosols in continental sequences might constitute marker-horizons with a greater precision. In fact, in the present state-of-the-art this method is not yet established and paleosols need to be defined more accurately. Retallack (1984) has, for example, studied the average rate of sedimentation in many Eo-Oligocene alluvial formations of South Dakota. For the purpose of evaluation he used the estimated time of formation of paleosols and geochronological and paleomagnetic measurements. Estimations made on the basis of paleosols were twenty times greater than the absolute ages (approximately 50 cm/1000 years against 2.5 cm/1000 years). The results are not erroneous, however, because from one formation to another there is a good correlation of the values on both scales.

A method presently applied to Quaternary paleosols (Mellor, 1985) allows better estimation of the degree of evolution and therefore the duration of a paleosol in ancient sequences. This is the method of *chronofunctions* or *chronosequences,* which proposes that certain parameters of profiles of alteration be dealt with separately as a function of time. These multiple parameters, left to the investigator's choice, are:
— thickness of profiles;
— relative thickness of certain horizons, the ratios of thicknesses;
— percentage of clays and organic matter, geochemistry etc.

In most cases chronofunctions may be expressed mathematically as:

$$Y = A \log T + B$$

where T = time and A and B are two constants.

This function asserts that *soils normally evolve towards a state of equilibrium.* Fig. 69 presents a theoretical approach to the parameters measuring the thickness of the diverse horizons in paleosols.

The method of chronofunction is a novel approach to paleosols. It enables better serialisation of the factor '**time of formation of a paleosol**' and consequently a better utilisation of paleosols as indicators of duration.

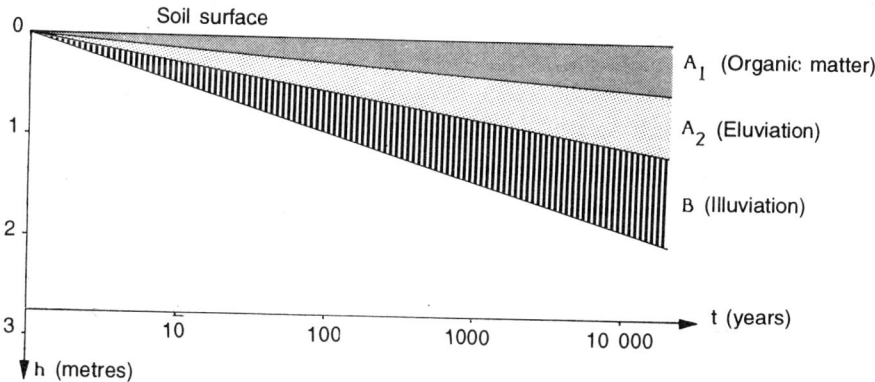

Fig. 69. *Theoretical differentiation of a pedological profile as a function of time (t). An approximation of thickness of each horizon is represented by the equation $H = A \log T$, but the scale is only indicative, as many factors influence the constant A.*

6.3.2 Paleoalterites as Stratigraphic Markers

At present, chronostratigraphic precision of continental sedimentary formations is much less compared to marine sedimentary formations. This is essentially due to the poor preservation of organic remains in continental facies. Under these circumstances, a paleoalterite of considerable extension may become a *stratigraphic marker* which cannot be overlooked, even though it may be only of *lithostratigrapic* value. It is uncontestable, for example, that the silcrete described in Sec. 2.5.3, and which occurs almost throughout the Coimbra-Leiria basin, evidences a significant event and consequently constitutes a lithostratigraphic marker. Attempts have been made to utilise such alterites in various other regions.

— Buurman (1972) associated a dozen paleosols with the peneplanation at Condroz (Belgium), which he considered as marker horizons from the Eocene to the present day.

— Mabesoone and Lobo (1978) proposed the use of four paleosols to interpolate the stratigraphy in north-eastern Brazil since the Eocene; these paleosols are of lateritic nature and can be correlated.

Paleoalterites found in Perigord (Daugas and Meyer, 1982) and at the northern border of the Massif Central (Thiry and Turland, 1984) record two successive evolutionary phenomena under the influence of two different climates: the first, a ferrallitic alteration, was followed by the formation of a siliceous cuirasse. The evolution of the Eocene climate may be derived from these observations, which changed from warm and humid conditions to hot and arid. The probability for the formation of such a *polyphased alterite* is so little compared to any other paleosol, that its stratigraphic importance increases significantly.

However, paleosols and paleoalterites must be used with great discretion as stratigraphic markers. **Violet Zones** (Sec. 2.5.3) of the Lower Triassic are revealing on the subject. They occur from Germany up to Portugal, passing through north-eastern France, the periphery of the Massif Central, Provence and Catalonia. In all these regions the facies of the paleosols are surprisingly similar. Some authors (Ortlam, 1971) have considered them stratigraphic markers but **palynological studies**, which are becoming increasingly important in continental sequences, have shown that no *correlation can be established* among violet zones; they developed in separate epochs whenever similar paleoenvironments prevailed (Durand and Meyer, 1982).

A good modern approach is to model the geometry of paleosurfaces using a computer. The comparison in the same area of paleosurfaces different in age is a very promising method (Wyns, 1991).

6.4 Knowledge of Paleogeography and Basin Dynamics

The knowledge gained from earlier descriptions of the duration and paleoclimates, naturally conveys information on paleogeogaphy; only complementary information will be dealt with here.

6.4.1 Problem of Fossilisation

Millot (pers. comm.) suggests that the *least soluble mineral* always takes precedence over other minerals in the environment under consideration. This principle is applicable to paleoalterites and it would appear evident that the mineral or minerals which were formed in the last phase of alteration, would be the only evidence of fossilisation.

If the alterite is not buried immediately after its formation, it may not fossilise unless the **climate evolves from maximum humidity to maximum aridity**; under opposite conditions, mechanical **erosion** causes the alterite to disappear. For example, in ancient sequences it is normally observed that ferruginous cuirasses silicify later in a less hydrolising climate.

It should thus be kept in mind that the present-day paleoalterites are *not representatives* of all the forms of alterites that can exist and hence the reconstruction of paleogeography on this premise is fairly complex.

6.4.2 Geochemical Paleoenvironments

It is necessary to differentiate between the chemistry of a formation and the chemistry of the surficial environments in which it could undergo weathering. *Chemical elements can play a very important role before elimination;* this is one of the merits of paleoalterites, that they provide evidence on the nature of the paleochemistry. This is particularly evident in the horizons of accumulation (Fig. 68), which present the total range of solutions from the most concentrated and basic (**arid climate**) to the most unsaturated (**very humid climate**).

Even if reconstruction of the profile proves to be almost impossible, partial observations provide very valuable information: for example, a **horizon** cannot be rubefied unless the interstitial waters are virtually free from bicarbonates. Similarly, a decolorised horizon must have the iron oxides removed in acidic and reducing solutions.

It is also possible to visualise a *hydric paleoregime* by virtue of certain features of paleoalterites.

— Permanent hydromorphy is evidenced by a totally reduced horizon (gley). This feature is not very characteristic of paleoalterites.

— Temporary hydromorphy is affected by fluctuations in the water table or by **marmorisation**. It characterises the zones of alternating *inundation and exundation*, which are very common in paleosols.

— The absence of traces of hydromorphy primarily indicates good **drainage**. Many types of soils cannot develop except in a completely exundated environment (see Table 5). Good drainage which favours the development of certain types of paleosols, must be proved for paleogeographic reconstructions. It implies *special types of relief* in the landscape and correlatively the existence of **pedological toposequences**.

Finally, the two most important geochemical parameters characteristic of a paleoenvironment are **Eh** and **pH**. Fig. 70 depicts an interpretation of alteration paleoprofiles oriented towards diverse possibilities.

6.4.3 Knowledge of Tectonic Activity

In-situ paleoalterites constitute layers *very sensitive* to vertical adjustments: positive movements lead to the erosion of alterites, while subsidence causes their submersion and restricts evolution. Analysis of paleoalterites provides data complementary to that derived from structural and sedimentological analysis of the basin. A few areas in which interpretation could well be valid are as follows:

— The *chronofunctions* introduced earlier (Sec. 6.3.1) not only measure the time, but when applied systematically to the profiles of weathering at the scale of the **basin**, make it possible to calculate the gradients and zones of particular types of evolution.

— Fig. 59 shows that the relationship between paleoalterites and *fluvial networks* changes in the tectonic context. There are also *alluvial sequences without paleosols*; these are more often deposited in regions of either very high subsidence or very weak subsidence (Sec. 3.1.5).

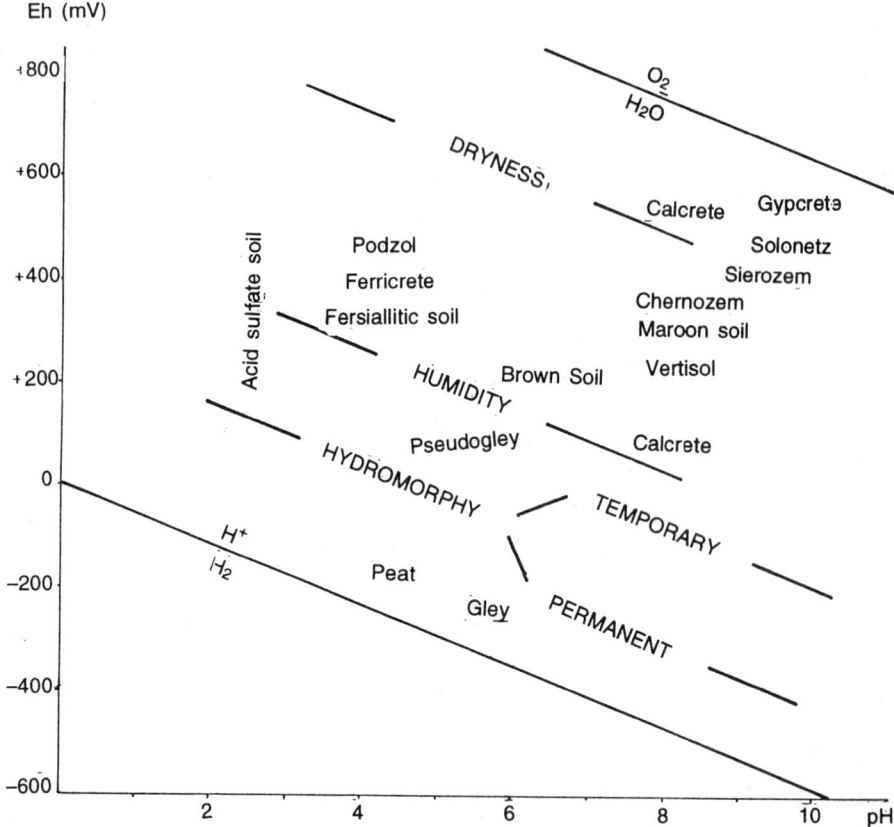

Fig. 70. *Theoretical stability field in Eh-pH diagram for principal type of paleosols. This diagram has only indicative value; in fact pH as well as redox potential change from one pedological horizon to another and also between two soils of the same denomination.*

— To reconstruct a sedimentary basin, the geologist sometimes has to determine the *'paleostrandline'*. More often the paleostrand is an aspect of marine studies; when marine features, especially paleontological ones disappear, this is taken to be the limit of the strandline. Systematic studies of continental sediments must be taken up as a complement to the paleogeographic scheme. The presence of *paleosols*, their *degree of evolution and regional distribution* are the effective elements for deciphering the existence or non-existence of marine influences. New results in this field are very interesting (Simon-Coicon et al., 1995).

— Analysis of the *events signifying the end of a* **sedimentary cycle** depends on the characteristics of the paleosurface which undergoes subsequent transgression. A study of this paleosurface and associated paleoalterites is thus of paramount importance for understanding the two superposed cycles.

— A *foreland* basin with fine stratigraphic texture could become a good *indicator of orogenic movements*, in fact more accurate than the mountain chain itself (Cavelier and

Pomerol, 1979). Investigations of the paleosurfaces and paleosols in such a complex provide complete information.

— It is necessary to reiterate that the differentiated paleosols indicate the top of the series deformed by tectonics because the profiles present a non-symmetric **differentiation.**

6.5 Appraisal

Paleoalterites and paleosols are a repository of knowledge. Whenever possible they should be utilised *together* with other methods. This permits determination of *certain parameters* for a better understanding of the *evolution of others*. For example, if the paleoclimate prevalent during the deposition of a formation is defined by means of paleontological criteria, the paleosol in turn indicates the duration relatively accurately.

Negative points: Information not recorded in paleoalterites
— *Climatic thrusts:* only the final result is recorded.
— Climatic evolution from relatively dry to more humid climates.
— *Permanent hydromorphy* causes chemical reduction of the sediments to the same extent as lacustrine or marine environments.
— Geologically *significant duration* for rapidly developed facies as marmorised horizons with root traces.
— Lithostratigraphic information when the sedimentary context is not clear.

Positive points: Information often recorded in paleoalterites
— Climatic evolution from relatively humid to arid climates.
— Traces of temporary *hydromorphy*.
— Traces of *good drainage* implying a certain paleorelief.
— Paleoclimatic limits.
— Large *paleoclimatic zones* for which the observation points are sufficient to describe the paleorelief and paleotoposequence.
— *Climatic drainage* and concentration of paleosolutions in horizons of accumulation.
— Indication of the type of fluvial network, deduced from the absence, presence and location of paleoalterites within sedimentary formations.
— Estimation of the *time necessary* for the development of a given paleosol.
— Position of the coastline (*paleostrand*) indirectly from the ensemble of paleosols.
— Conditions at the commencement and termination of sedimentary cycles, these limits being signified by paleosurfaces.

7. GENERAL CONCLUSIONS

Paleosols, the best organised and the easiest to interpret among paleoalterites, may be investigated with a fair amount of rationality in the present state-of-the-art. Even if this cannot revolutionise our knowledge of the paleoenvironments, it definitely does contribute to further precision of them.

Among the methods which make valuable contributions to the reconstruction of paleoenvironments, paleoalterology is comparable to some sophisticated analytical methods which are more expensive and cumbersome. In contrast paleoalterology is more competitive financially. There is one weakness in comparing analytical methods, however; the human factor plays a very important role in paleoalterology, thereby giving the impression of a traditional discipline that attaches more importance to observations, the effectiveness of which is sometimes dubious. In the context of present-day technology there could be a loss of credibility should the geologist fail to exercise good judgement: an intuitive sense while making observations in the field, a discerning capability, a taste for modern methods of investigation and last but not least, circumspection and a balanced interpretation.

Knowledge in paleoalterology needs to be expanded to keep pace with the progress made in other disciplines of geology. The first step towards solving problems which remain unsolved is to properly outline the problem; the following procedure is quite safe:

— Petrographic particularities of continental sedimentary formations indicate that these evidently traverse a different path of diagenesis than those of marine formations.

— Definition of coastlines in paleogeographic reconstructions.

— Distinction between meteoric phenomena and the phenomena of hydrothermal transformations and diagenesis at the scale of paleosurfaces related to discordances.

Probably in the near future a large number of paleoalterites will be discovered and interpreted, but it is desirable that this quantitative increase be accompanied by a matching refinement in the methods of investigation. This involves advancement in the technics used in all the disciplines of earth science. However, such refinement also involves a new conceptualisation. Like paleontologists, who have established the classification of biological forms that no longer exist, paleopedologists must envisage a classification of paleosols which, without completely neglecting existing models, should not systematically and solely refer to them.

8. GLOSSARY

8.1 Some Important Soil Types

The following definitions were adapted from Duchaufour (1976, 1977, 1984). Emphasis has been given to the morphological and mineralogical characteristics which are likely to be found in fossil soils. It should be noted in particular that surficial horizons rich in organic matter ($A_0 - A_1$) exist in each of the soil types described below, but less significance has been accorded to them because of their limited geological importance.

ACID BROWN SOIL (0.2 to 1 m in thickness): brown soil developed on acid parent rock; the iron liberated by primary minerals obstructs leaching; it is characteristic of deciduous forests in temperate climates.

ALLUVIAL SOIL (0.5 to 1.5 m in thickness): little evolved soil, develops on loose alluvial material; hydromorphy could be permanent or temporary, leaving imprints; weathering is less in temperate climates, increasing in warm and humid climate.

BROWN SOIL (0.2 to 1 m in thickness): differentiated soil underneath a surficial humic horizon, an alteration horizon of brown colour and polyhedral structure; it characterises temperate or cold climates well.

CHESTNUT SOIL (0.8 to 2 m in thickness): intermediate between chernozems and sierozems; characteristic of dry climate of bare steppes.

CHERNOZEM (1 to 2 m in thickness): little differentiated soil, uniformly black due to the incorporation of profuse organic matter; characteristic of dense steppe in temperate continental climate.

FERRALLITE (several metres in thickness): most evolved form of ferrallitic soils.

FERRALLITIC SOIL (several metres in thickness): thick evolved soil characterised by almost complete weathering of primary minerals; cations are rendered soluble and eliminated from the profile so that only the oxides of iron, aluminium and possibly part of the silica remain; red tones; characteristic mineral either kaolinite (monosiallitisation) or gibbsite (allitisation); these soils represent well-drained environments in warm and humid climate.

FERSIALLITIC SOIL (1 to 2 m in thickness): coloured by iron hydroxide wherever primary minerals underwent moderate weathering; profile slightly differentiated (brown fersiallitic soil to simple weathering horizon) or more evolved (leached fersiallitic soil to argillic B_t horizon); clay minerals are mainly 2/1 type (bisiallitisation); the rubefaction which characterises these soils presupposes leaching of carbonates from the parent rock and a warm to markedly dry climate.

GLEY (0.5 m to tens of metres in thickness): little differentiated soil to which reduced iron imparts a grey or somewhat dark colour; formed rapidly under the influence of reducing underground water; permanent hydromorphy independent of climate.

LEACHED BROWN SOIL (0.5 to 1.5 m in thickness): brown soil which has undergone 'leaching' or mechanical extraction of clay and iron; a depleted horizon, distinctly A_2, overlies a B_t horizon with argillic coloration; this evolution takes place in temperate climates when the parent rock is free of carbonates or has been previously leached.

MAROON SOIL (0.5 to 1.5 m in thickness): quite uniformly reddish; moderate weathering of primary minerals, often developing a calcic or petrocalcic horizon (calcareous crust) towards the base of the profile; characteristic soil of bushy vegetation in Mediterranean or subtropical climates.

PLANOSOL (0.5 to 1.5 m in thickness): soil in which the surficial horizon poor in clay and iron (albic horizon) overlies an almost impermeable argillic horizon; formed by temporary hydromorphy in highly contrasting warm climates.

PODZOL (0.5 to 1 m in thickness): well-differentiated soil; horizon A_1 is light grey in colour, uniquely residual quartz while horizon B is composed of two horizons: B_h—black with organic matter and B_s—rust-coloured, rich in oxides (often concretionary); develops favourably on acid parent rocks; most often characterises boreal or cold temperate climates.

PODZOLIC OCHREOUS SOIL (0.3 to 0.6 m in thickness): similar to podzols but less evolved; horizon A_2 is made up of grey patches; horizons B_h and B_s cannot be distinguished from each other; poor evolution due to young age of the soil or less acidic nature of the parent rock; common in temperate climate.

PSEUDOGLEY (0.5 to 1.5 m in thickness): iron is mobilised by temporary hydromorphy, which causes the appearance of ochreous or rusty patches in decolorised material (marmorisation, glosses); hydromorphy is most often due to an impermeable horizon at the base of the profile; develops mainly in temperate climate but is not characteristic of it.

RENDZINA (0.2 to 0.6 m in thickness): little differentiated soil which develops on a soft parent rock, limestone or dolomite; abundant carbonate-initiated alteration leads to formation of a surficial horizon with sufficiently stable gritty structure; characteristic of temperate climate.

SIEROZEM (0.5 to 2 m in thickness): little differentiated soil characteristic of sparse steppe in arid climate; carbonates and sulfates may precipitate at depth.

SOLONETZ (0.5 to 1 m in thickness): rich in Na^+ ion and derived from weathering of sodic minerals (albite); sodic clays are leached and concentrated in the natric horizon; pH value often high and environment aggressive for silicates; soil of arid climate.

TROPICAL BROWN SOIL (0.5 to 1.5 m in thickness): little evolved soil with a horizon of cambic alteration; underwent initial rubefaction in contrasting tropical climates.

TROPICAL FERRUGINOUS SOIL (1 to 3 m in thickness): similar to ferrallitic soil; weathering of primary minerals incomplete because of less humid climate; red tones.

VERTISOL (1 to 2 m in thickness): develops on a parent rock rich in swelling clays; alternate swelling and retraction results in strong mechanical stresses in the interior of the profile, with the appearance of almost oblique frictional surfaces (slickensides); these soils develop locally where drainage is poor; characteristic of warm climates and highly contrasting seasons; waterlogged hydromorphy followed by strong desiccation.

8.2 Some Fundamental Pedological Micromorphology

The terms proposed below are adapted from Brewer (1964).

8.2.1 Transition from Macrostructure to Microstructure

At the macroscopic scale the structures of the soil and the pedological horizons vary as a function of the parent rock and of biological activity. These structures are massive, polyhedric, prismatic (vertically elongated elements), plaquettes, nuciform, gritty etc. At a

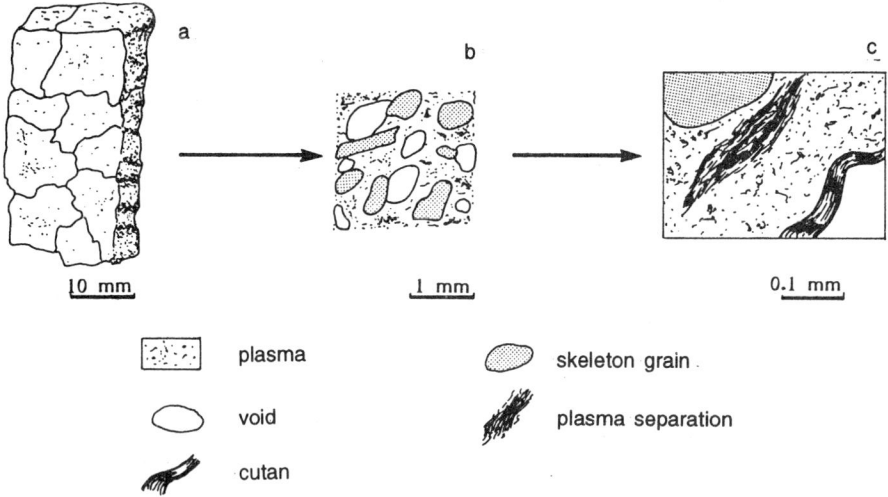

Fig. 71. Organisation of the soil at diverse scales.

a) Large polyhedrals composed of several elementary polyhedrals.

b) Details of an elementary polyhderal showing the fabric of fundamental constituents: skeleton grains, plasma and void.

c) At this scale plasma separation can be observed under the microscope in polarised light; clays have a common orientation in the plasma (insepic separation); cutan is an example of a pedological trait.

finer scale, microstructures define the relationship that exists between the fundamental constituents of the soil (Fig. 71).

8.2.2 Fundamental Constituents

PLASMA: every kind of material, mineral or organic, which is of colloidal size, less than 2 μm.

S-MATRIX: material in which pedogenesis occurs; it is constituted of plasma, skeleton grains and voids.

SKELETON GRAINS: mineral or organic elements greater than 2 μm in size.

VOIDS: cavities of various types such as chambers, pedochannels or fissures.

8.2.3 Basic Fabric

Based on the relation between skeleton grains and plasma, several types of fabrics can be distinguished. These essentially indicate the phenomena of eluviation (particularly granular fabric) or illuviation (porphyritic fabric).

AGGLOMEROPLASMIC FABRIC: plasma occurs as a weak filling between skeleton grains (Fig. 72b).

GRANULAR FABRIC: where plasma is absent (Fig. 72d).

INTERTEXTIC FABRIC: skeleton grains connected by plasma bridges (Fig. 72c).

PORPHYROSKELIC FABRIC: skeleton grains embedded in plasma-like phenocrysts in a crystalline rock with porphyritic texture (Fig. 72a).

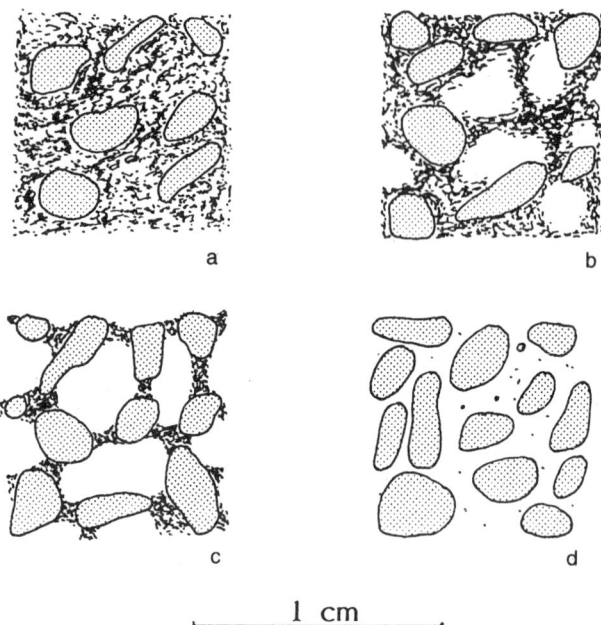

Fig. 72. Elementary fabric of a soil (figures are those of Fig. 71).
  a) Porphyritic: skeleton grains embedded in plasma.
  b) Agglomeratic: discontinuous plasma in-between the grains.
  c) Intertextic: plasma bridge between the grains.
  d) Granular: absence of plasma.

8.2.4 Plasma Separation

The plasma constituents, especially phyllitic minerals, tend to acquire a common orientation in certain zones of the S-matrix. These zones are birefractive under an optical microscope and known as sepic plasmic fabric (the word *sepic* is derived from *separation*). These separations are the result of physical stress during pedogenesis, for example, bioturbation, or alternate humidification-desiccation.

INSEPIC FABRIC (Latin '*insula*': island): some isolated veins develop in the assemblage of clay minerals, which have a common orientation (see Fig. 71c).

LATTISEPIC FABRIC (from '*lattice*'): variation of the bimasepic fabric where the two directions of orientation of phyllites are perpendicular to each other (Fig. 73f).

MASEPIC FABRIC (from '*matrix*'): clays are oriented along a preferred direction valid throughout the sample (Fig. 73d). The term **BIMASEPIC** is used when there are two directions of orientation.

MOSEPIC FABRIC (from '*mosaic*'): extreme development of insepic fabric; veins are intrusive throughout the plasma but with no preferred orientation throughout the rock (Fig. 73a).

SKELSEPIC FABRIC (from '*skeleton grains*'): phyllites are oriented parallel to the surfaces of skeleton grains (Fig. 73c).

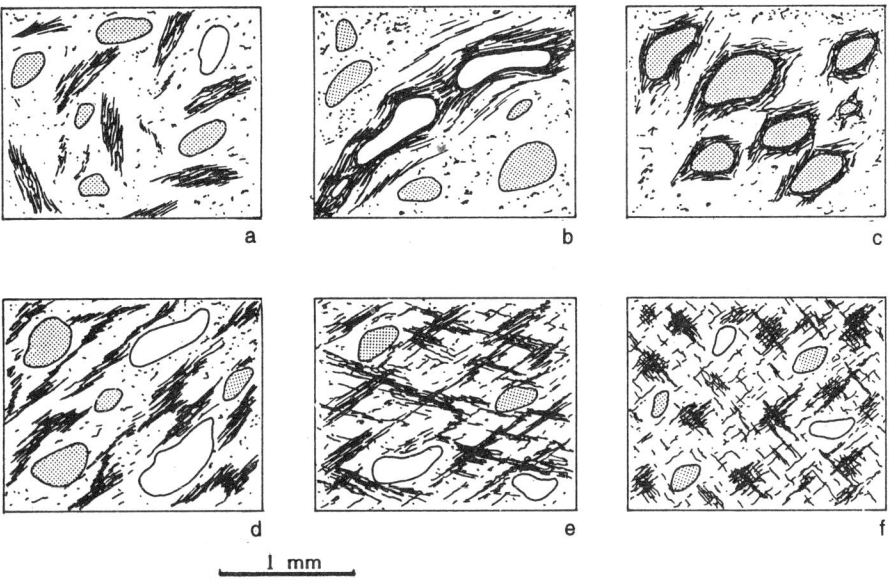

Fig. 73. *Principal types of plasma separation (figures those of Fig. 71)*
 a) *Mosepic separation: a number of independent bands.*
 b) *Vosepic separation: phyllites oriented around void walls.*
 c) *Skelsepic separation: phyllites around skeleton grains.*
 d) *Masepic separation: unique orientation.*
 e) *Bimasepic separation: two preferred directions of orientation.*
 f) *Lattisepic separation: two perpendicular directions of orientation.*

VOSEPIC FABRIC (from '*void*'): phyllites are oriented parallel to the walls of the voids (Fig. 73b).

8.2.5 Pedological Features

These are the specific features locally observed in the soil; they are of several types.

CUTAN (Latin '*cutis*': skin): concentration of one of the constituents of the soil on a natural surface; three types of cutans are distinguished based on the origin and processes of development:

— *Diffusion cutans:* certain elements of the plasma migrate by chromatography towards the surface of the polyhedrals (iron hydroxides, calcite); this mainly indicates alternate humidification-desiccation.

— *Illuviation cutans:* water circulating through pedochannels carries particles or dissolved elements which are deposited at the surface of voids.

— *Stress cutans:* stresses in the soil modify the plasma (orientation of clays parallel to the frictional surfaces (Fig. 74a)); stress cutans which coat the frictional surfaces, are more or less curved and visible to the naked eye, are called slickensides.

Cutans are classified according to their mineralogy and chemistry:

— **Argillan:** surficial deposit of clay minerals, clear yellow under a microscope (Fig. 6a).

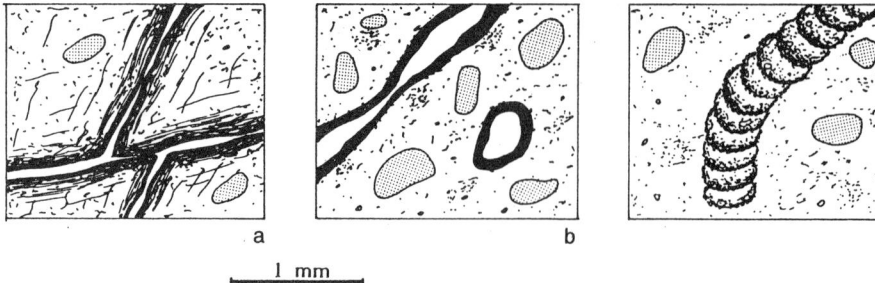

Fig. 74. Examples of pedological traits (figures those of Fig. 71).
a) Stress cutans developed along the fissures.
b) Opaque ferrans coating the walls of voids.
c) Striotubules caused by burrowing animals.

— **Ferri-argillan:** surficial deposit of clay minerals intimately mixed with oxides and hydroxides of iron; orange under a microscope.
— **Calcitan:** coating of calcite crystals.
— **Ferran:** surficial deposit essentially of oxides or hydroxides of iron; generally opaque (Fig. 74b).
— **Gibbsan:** surficial deposit of gibbsite (see Fig. 51)
— **Mangan:** surficial deposit of manganese oxides (see Fig. 47).
— **Silan:** surficial deposit of silica minerals (see Fig. 28).
— **Siltan:** fine laminar coating in which silt-sized grains are identifiable.
GLAEBULES (Latin '*glaebula*': small mounds of earth): small mounds of millimetric or centimetric size, more or less rounded, and with constituents that differ from those of the surrounding matrix.
PAPULES (Latin '*papula*': pimple): aggregates of oriented clay minerals formed by the destruction and reworking of argillans or ferri-argillans (see Fig. 6b) within the soil.
PEDOTUBULES: tubular cavities with diameters ranging from a few millimetres to a few centimetres, filled by material different from the S-matrix. The most remarkable are the **STRIOTUBULES** of which the fillings are a succession of quite characteristic menisci (Fig. 74c). Striotubules are burrows, especially of lombricides, whereas irregularly filled pedotubules are the abandoned cavities caused by roots or animals and later filled by sediments.

REFERENCES

Abed A.M. (1979). — Lower Jurassic lateritic red beds from Central Arabia, *Sedim. Geol.*, 24, 1/2, pp. 149–156.
Adolphe J.P. (1981). — Exemples de contribution microorganique dans les construction carbonatées continentales. *Association Française Karstologie, Mém.* 3, pp. 15–30.
Alidou S., Lang J., Lucas G. (1977). — Sur un rôle possible joué par les Termites dans la fossilisation des restes végétaux. *Sci Géol., Bull.*, 30, 3, pp. 203–206.
Allen J.R.L. (1964). — Studies in fluviatile sedimentation: six cyclothems from the Lower Old Red Sandstone, Anglo-Welsh Basin. *Sedimentology*, 3, pp. 163–198.
Allen J.R.L. (1974). — Studies in fluviatile sedimentation: implications of pedogenic carbonate units, Lower Old Red Sandstone, Anglo-Welsh outcrop. *Geol. J.*, 9, 2, pp. 181–208.
Allen Curran H. (1985). — Biogenic structures: their use in interpreting depositional environments. SEPM Spec. publ., 35, 347 p.
Anadon P., Zamarreño I. (1981). — Paleogenè nonmarine algal deposits of the Ebro Basin, Northeastern Spain. *In:* Phanerozoic Stromatolites, C. Monty éd., Springer-Verlag, Berlin, pp. 140–154.
Arakel A.V. (1982). — Genesis of calcrete in Quaternary soil profiles. Hutt and Leeman Lagoons, Western Australia. *Jour. Sedim. Petrol.*, 52, 1, pp. 109–125.
Arakel A.V., McConchie D. (1982). — Classification and genesis of calcrete and gypcrete lithofacies in paleodrainage systems of inland Australia and their relationship to carnotite mineralization. *Jour. Sedim. Petrol.*, 52, 4, pp. 1149–1170.
Arbey F. (1980). — Les formes de la silice et l'identification des évaporites dans les formations silicifiées. *Bull. Centres Rech. Explor.-Prod. Elf-Aquitaine*, 4, pp. 309–365.
Baltzer F. (1970). — Étude sédimentologique du Marais de Mara et des formations quaternaires voisines. Expédition française sur les récifs coralliens de la Nouvelle Calédonie. Vol. 4. Fondation Singer-Polignac éd., Paris, 142 p.
Barbier J. (1978). — A propos de calcrètes, d'érosion et de la répartition des gîtes d'uranium intragranitiques français. *B.R.G.M. Bull.*, section II, I, pp. 31–38.
Bardossy G. (1981). — Les bauxites européennes, leur géologie, prospection et valorisation économique. *Chronique Rech. Min.*, 459, pp. 5–21.
Barriere J. (1971). — Limites d'utilisation des paléosols pour la reconstitution de paléoclimats. *C.R. Acad. Sci. Paris*, 272 D. pp. 2426–2429.
Bartoli F., Meyer R., Moura F., Samama J.C. (1983). — Caractérisation chimicominéralogique de l'opale noble des gisements du nord-est du Brésil. *C.R. Acad. Sci. Paris*, série II, 296, pp. 625–630.
Battiau-Queney Y. (1984). — Désilicification des grès quartzites du Millstone Grit du Pays-de-Galles et paysages associés, Réunion R.C.P. 706 « Paléoaltérations et paysages associés », Paris, 20 nov. 1984, 2 p.
Beauchamp J. (1981). — Structure et mode de silicification de quelques bois fossiles. *Sci. Géol., Bull.*, 34, I, pp. 13–20.
Bech J., Nahon D., Paquet H., Ruellan A., Millot G. (1980). — Sur l'extension géographique et climatique des phénomènes d'épigénie par la calcite dans les encroûtements calcaires. Exemple de la Catalogne. *C.R. Acad. Sci. Paris*, 291 D, pp. 371–376.
Becker A. (1992). — A time-space model for the genesis of Early Tertiary laterites from the Jos Plateau, Nigeria. *J. African Earth Sci.*, 15, pp. 265–269.
Bernard A. (1972). — Le karst et les métallisations sulfurées. Conférences et séminaires recyclage métallogénie. E.N.S.G. Nancy, t.I. chap. 2, 54 p.

Bertrand J.P., Jelisejeff A. (1974). — Formation d'évaporites par des processus d'évaporation capillaire. *Rev. Géogr. phys. géol. dynam.* (2) 16, 2, pp. 161–170.
Bertrand-Sarfati J., Moussine-Pouchkine A. (1983). — Pedogenic and diagenetic fabrics in the Upper Proterozoic Sarnyéré Formation (Gourma, Mali). *Precambrian Research,* 20, pp. 225–242.
Bignot G. (1995). — Les deux épisodes à Microcodium du Paléogène parisien replacés dans un contexte péritéthysien. *Newsl. Stratigr.* 32 (2), pp. 79–89.
Billaud Y. (1982). — Les paragenèses phosphatées du paléokarst des phosphorites du Quercy. Thèse, Lyon, 135 p.
Blanco J.A., Cantano M. (1983). — Silicification contemporaine à la sédimentation dans l'unité basale du Paléogène du Bassin du Duero (Espagne). *Sci. Géol., Mém.,* 72, pp. 7–18.
Bocquier G. (1971). — Genèse et évolution de deux toposéquences de sols tropicaux du Tchad. Interprétation biogéodynamique. Thèse Sci., Strasbourg et ORSTOM, Paris, 364 p.
Bodergat A.M. (1974). — Les microcodiums. Milieux et modes de développement. *Doc. Lab. Géol. Fac. Sci. Lyon,* 62, pp. 137–235.
Boucot A.J., Dewey J.F., Dineley D.L., Fletcher R., Fyson W.K., Griffin J.G., Hickox C.F., Mckerrow W.S., Ziegler A.M. (1974). — Geology of the Arisaig Area, Nova Scotia. *Geol. Soc. Am. Special Paper,* 139, 191 p.
Boulange B., Bocquier G. (1983). — Le rôle du fer dans la formation des pisolites alumineux au sein des cuirasses bauxitiques latéritiques. *Sc. Géol. Mém.* 72, pp. 29–36.
Brewer R. (1964). — Fabric and mineral analysis of soils. Wiley, New York, 470 p.
Brewer R., Haldane A.D. (1957). — Preliminary experiments in the development of clay orientation in soils. *Soil Sci.,* 84, pp. 301–309.
Brinkman R., Jongmans A.G., Miedema R., Maaskant P. (1973). — Clay decomposition in seasonally wet, acid soils: micromorphological, chemical and mineralogical evidence from individual argillans. *Geoderma,* 10, pp. 259–270.
Briot Ph., Fuchs Y. (1978). — The importance of paleoclimatic and morphological factors in the genesis of certain uranium-bearing calcretes. 10th Int. Congress Sedim., Jerusalem, 11 p. dactyl. et abstract p. 91.
Brophy G.P., Scott E.S., Snellgrove R.A. (1962). — Sulfate studies II. Solid solution between alunite and jarosite. *Am. Mineralogist,* 47, pp. 112–126.
Buol S.W., Hole F.D., McCracken R.J. (1973). — Soil genesis and classification. Iowa State Univ. Press, Ames, 360 p.
Buurman P. (1972). — Paleopedology and stratigraphy on the Condrusian peneplain (Belgium) with a reconstitution of a paleosol. Centre for Agricultural Publishing and Documentation, Wageningen, 67 p.
Buurman P. (1980). — Paleosols in the Reading Beds (Paleocene) of Alum Bay, Isle of Wight, U.K. *Sedimentology,* 27, pp. 593–606.
Buurman P., Breemen N. van, Henstra S. (1973). — Recent silicification of plant remains in acid sulphate soils. *N. Jb. Miner. Mh.,* 1973, 3, pp. 117–124.
Calvet F., Julia R. (1983). — Pisoids in the caliche profiles of Tarragona (N.E. Spain). *In*: Coated Grains, T.M. Peryt, ed., Springer-Verlag, pp. 456–473.
Cantinolle P., Didier P., Meunier J.D., Parrron C., Guendon J.L., Bocquier G., Nahon D. (1984). — Kaolinites ferrifères et oxy-hydroxydes de fer et d'alumine dans les bauxites des Canonnettes (S.E. de la France). *Clay Minerals,* 19, pp. 125–135.
Carlisle D. (1978). — Characteristics and origins of uranium-bearing calcretes in Western Australia and South West Africa. Calcrete symposium., 10th Int. Congress Sedim., Jerusalem, p. 119.
Casanova J. (1981). — Morphologie et biolithogenèse des barrages de travertins. *Association Française Karstologie, Mém.* 3, pp. 45–54.
Cavelier C., Pomerol C. (1979). — Chronologie et interprétation des événements tectoniques cénozoïques dans le Bassin de Paris. *Bull. Soc. Géol. France,* (7), XXI, pp. 33–48.
Cayeux L. (1929). — Les roches sédimentaires de France. Roches siliceuses. Imprimerie nat., Paris, 773 p.

REFERENCES

Cerling T.E. (1984). — The stable isotopic composition of modern soil carbonate and its relationship to climate. *Earth Planet. Sci. Lett.*, 71, pp. 229-240.
Chafetz H.S. (1982). — The Upper Cretaceous Beartooth Sandstone of Southwestern New Mexico: a trangressive deltaic complex on silicified paleokarst. *Jour. Sedim. Petrol.*, 52, 1, pp. 157-169.
Chafetz H.S., Folk R.L. (1984). — Travertines: depositional morphology and the bacterially constructed constituents. *Jour. Sedim. Petrol.*, 54, 1, pp. 289-316.
Chesworth W. (1973). — The residual system of chemical weathering: a model for the chemical breakdown of silicate rocks at the surface of the earth. *J. Soil Sci.*, 24, pp. 69-81.
Chesworth W., Dejou J., Larroque P. (1981). — The weathering of basalt and relative mobilities of the major elements at Belbex, France. *Geochim. Cosmochim. Acta*, vol. 45, pp. 1235-1243.
Chesworth W., Dejou J., Kimpe C. de, Macias Vasquez F., Cantagrel J.M., Larroque P., Garcia Paz C., Garcia Rodeja E. (1983). — Importance de la fersiallitisation sur les basaltes miocènes du Massif Central. Principales caractéristiques de cette pédogenèse. *Sci. Geol., Mém.*, 73, pp. 53-62.
Cohen A.S. (1982). — Paleoenvironments of root casts from the Koobi Fora Formation, Kenya. *Journ. Sedim. Petrol.*, 52, 2, pp. 401-414.
Combes P.J. (1978a) — Nouvelles données sur les relations entre la paléogéographie et la gîtologie des bauxites du troisième horizon dans la zone du Parnasse (Grèce). 4th Int. Congress for study of bauxites, Athens, vol. 1, pp. 92-100.
Combes P.J. (1978b). — Karst précoce et karst secondaire du troisième horizon de bauxite dans la zone du Parnasse (Grèce). 4th. Int. Congress for study of bauxites, Athens, vol. 1, pp. 101-113.
Combes P.J. (1984). — Regards sur la géologie des bauxites: aspects récents sur la genèse de quelques gisements à substratum carbonaté. *Bull. Centres Rech. Explor.-Prod. Elf-Aquitaine*, 8, 1, pp. 251-274.
Conrad G. (1969). — L'évolution continentale posthercynienne du Sahara algérien. C.N.R.S., Paris, 527 p.
Cortelezzi C.R., Kilmurray J.O. (1965). — Surface properties and epigenetic fractures of gravels from Patagonia, Argentina. *Jour. Sedim. Petrol.*, 35, pp. 976-980.
Crouzel F., Meyer R. (1975). — Encroûtements calcaires dans l'Oligo-Miocène du Bassin d'Aquitaine. *C.R. Soc. Géol. France*, pp. 112-114.
Crouzel F., Meyer R. (1977a). — Les calcaires lacustres du Miocène Aquitain. Participation des lacs à l'édification d'un ensemble molassique continental. *Sci. Terre*, XXI, 3, pp. 237-250.
Crouzel F., Meyer R. (1977b). — Néoformation de dolomite dans des encroûtements pédologiques de l'Oligo-Miocène aquitain. 5e *Réunion an. Sci. Terre*, Rennes, p. 177.
Crouzel F., Meyer R. (1983). — Faciès silicifiés d'origine météorique dans le Miocène continental de l'Armagnac. *Bull. Soc. Géol. France*, (7), 25, pp. 19-23.
Daugas F. (1981). — Les dépôts continentaux du Lias inférieur, du Tertiaire et du Quaternaire ancien au Nord du Périgord. Sédimentation, paléoaltération et stratigraphie. Thèse, Nancy, 153 p.
Daugas F., Meyer R. (1982). — Les dépôts continentaux cénozoïques au Nord du Périgord. Place des faciès « sidérolithiques ». *C.R. Acad. Sci. Paris*, série II, t. 295, pp. 493-496.
Davaine J.J. (1980). — Les croûtes silicofluorées du Bazois (Morvan). Thèse Nancy et *Mém. B.R.G.M.* 104, pp. 209-341.
Dejou J., Chesworth W., Larroque P. (1982). — Données nouvelles sur l'évolution superficielle fersiallitique subie par les basaltes pontiens du Bassin d'Aurillac (Cantal, France). Cas du profil de Saint-Etienne de Carlat et considérations paléoclimatiques. *Pédologie*, 32, 1, pp. 67-83.
Didyk B.M., Simoneit B.R.T., Brassell S.C., Eglinton G. (1978). — Organic geochemical indicators of palaeoenvironmental conditions of sedimentation. *Nature*, 272, pp. 216-222.
Donsimoni M., Giot D. (1977). — Les calcaires concrétionnés lacustres de l'Oligocène supérieur et de l'Aquitanien de Limagne (Massif Central). *Bull. B.R.G.M.* (2), 1, 2, pp. 131-169.
Duchaufour Ph. (1976). — Atlas écologique des sols du monde. Masson, Paris, 178 p.
Duchaufour Ph. (1977). — Pédologie 1: Pédogenèse et classification. Masson, Paris, 477 p.
Duchaufour Ph. (1984). — Abrégé de pédologie. Masson, Paris, 220 p.
Ducloux J. (1973). — Etude micromorphologique des paléosols du haut niveau des alluvions du Lay à Mareuil (Vandée). *Bull. Ass. Fr. Étude Quaternaire*, 1973-3, pp. 131-149.

Durand J.H. (1959). — Les sols rouges et les croûtes en Algérie. Direction Hydraulique Equipement Rural, étude no. 7, Alger, 188 p.

Durand M. (1975). — Nature des colorations violettes et vertes de certains grès triasiques. *C.R. Acad. Sci. Paris,* 280 D, pp. 2737–2740.

Durand M. (1978). — Paléocourants et reconstitution paléogéographique. L'exemple du Buntsandstein des Vosges méridionales (Trias inférieur et moyen continental). *Sci. Terre,* XXII, 4, pp. 301–390.

Durand M., Meyer R. (1982). — Silicifications (silcrètes) et évaporites dans la Zone-limite violette du Trias inférieur lorrain. Comparaison avec le Buntsandstein de Provence et le Permien des Vosges. *Sci. Géol., Bull.,* 35, 1–2, pp. 17–39.

Ellenberger F., Feys, R., Trichet J. (1967). — Paléotopographie et podzols résiduels au sommet des Sables de Fontainebleau. *Cr. Acad. Sci. Paris,* 264 D, pp. 689–692.

Emberger L. (1968). — Les plantes fossiles dans leurs rapports avec les végétaux vivants. Masson, Paris, 758 p.

Engelhardt W. von (1977). — Sedimentary petrology, III : The origin of sediments and sedimentary rocks. Schweizerbart'sche Verlagsbuchhandlung, Stuttgart, 359 p.

Erhart H. (1956, 2e éd. 1967). — La genèse des sols en tant que phénomène géologique. Masson, Paris, 177 p.

Esteban M., Klappa C.F. (1983). — Subaerial exposure environment. *In*: Carbonate Depositional Environments, P.A. Scholle, D.G. Bebout, C.H. Moore, ed. *A.A.P.G. Mém.* 33, pp. 2–54.

Esteban M., Pomar Goma L., Marzo M., Anadon P. (1977). — Naturaleza del contacto entre el Muschelkalk inferior y el Muschelkalk medio de la zona de Aiguafreda. *Cuadernos Geologia Ibérica,* 4, pp. 201–210.

Esterle M. (1967). — Contribution à l'étude de la genèse et de l'évolution des bauxites karstiques de Provence. Géologie, minéralogie, géochimie des formations bauxitiques de Saint-Maximin, Ollières, Pourcieux (Var). Thèse, Paris, 122 p.

Even G. (1978). — Mise au point sur les connaissances actuelles concernant les tonsteins. Exemples d'utilisation stratigraphique de ces niveaux. 103e Congr. Nat. Soc. Sav., Sciences, IV, Nancy, pp. 315–322.

Even G., Samama J.C. (1969). — Argiles de socle et argiles de couverture: le problème des altérations et des néoformations au contact entre le Trias et le socle hercynien du Vivarais. *C.R. Acad. Sci. Paris,* 268 D, pp. 3005–3008.

Flach K.W. Nettelton W.D., Gile L.H. Cady J.G. (1969). — Pedocementation: induration by silica, carbonates, and sesquioxides in the Quaternary. *Soil Science,* 107, 6, pp. 442–453.

Folk R.L. (1971). — Caliche nodule composed of calcite rhombs. *In*: Carbonate Cements, R.K. Bricker, ed., Johns Hopkins Press, Baltimore, pp. 167–168.

Folk R.L., Pittman J.S. (1971). — Length-slow chalcedony: a new testament for vanished evaporites *J. Sedim. Petrol.,* 41, 4, pp. 1045–1058.

Freytet P. (1964). — Le Vitrollien des Corbières orientales: réflexions sur la sédimentation «lacustre» nord-pyrénéenne; divagation fluviatile, biorhexistasie, pédogenèse. *Rev. Géogr. Phys. Géol. Dynam.,* VI, 3, pp. 179–199.

Fretet P. (1970). — Les Dépôts continentaux et marins du Crétacé supérieur et des couches de passage à l'Eocène en Languedoc. Thèse Sci., Orsay, 534 p.

Freyet P. (1971). — Paléosols résiduels et paléosols alluviaux hydromorphes associés aux dépôts fluviatiles dans le Crétacé supérieur et l'Eocène basal du Languedoc. *Rev. Géogr. Phys. Géol. Dynam.,* XIII, 3, pp. 245–268.

Freytet P. (1973). — Petrography and paleo-environment of continental carbonate deposits with particular reference to the Upper Cretaceous and Lower Eocene of Languedoc (Southern France). *Sedim. Geol.,* 10, pp. 25–60.

Freytet P., Plaziat J.C. (1982). — Continental carbonate sedimentation and pedogenesis. Late Cretaceous and Early Tertiary of Southern France. Contributions to Sedimentology 12. Schweizerbart'sche Verlagsbuchhandlung, Stuttgart, 213 p.

FRITZ W.J. (1980a). — Reinterpretation of the depositional environment of the Yellowstone « fossil forests ». *Geology,* 8, pp. 309–313.

REFERENCES

Fritz W.J. (1980b). — Stumps transported and deposited upright by Mount St. Helens mud flows. *Geology*, 8, pp. 586–588.

Füchtbauer H. (1974). — Sedimentary petrology, II: Sediments and Sedimentary Rocks I. Schweizerbart'sche Verlagsbuchhandlung, Stuttgart, 464 p.

Gall J.C. (1971). — Faunes et paysages du Grès à Voltzia du Nord des Vosges. Essai paléoécologique sur le Buntsandstein supérieur. Thèse Sci., Strasbourg, et *Mém. Serv. Carte Géol. Als. Lor.*, 34, Strasbourg, 318 p.

Gay A.L., Grandstaff D.E. (1980). — Chemistry and mineralogy of Precambrian paleosols at Elliot Lake, Ontario, Canada. *Precambrian Research*, 12, pp. 349–373.

Ginsburg L. (1974). — Les Rhinocérotidés du Miocène de Sansan (Gers). *C.R. Acad. Sci. Paris*, 278 D, pp. 597–600.

Goldbery R. (1978). — Early diagenetic, nonhydrothermal Na-alunite in Jurassic flint clays, Makhtesh Ramon, Israel. *Geol. Soc. Am. Bull.*, 89 pp. 687–698.

Goldbery R. (1979). — Sedimentology of the Lower Jurassic flint clay bearing Mishor Formation Makhtesh Ramon, Israel, *Sedimentology*, 26, 2, pp. 229–251.

Goldbery R. (1982). — Paleosols of the Lower Jurassic Mishhor and Ardon Formations (« Laterite derivative facies »), Makhtesh Ramon, Israel, *Sedimentology*, 29, 5, pp. 669–690.

Goldbery R., Beyth M. (1984). — Lateritization and ground water alteration phenomena in the Triassic Budra Formation, south-western Sinai. *Sedimentology*, 31, pp. 575–594.

Goldstein R.H. (1991). — Stable isotope signatures associated with paleosols, Pennsylvanian Holder Formation, New Mexico. *Sedimentology*, 38 pp. 67–77.

Goudie A.S. (1973). — Duricrusts in Tropical and Subtropical Landscapes. Clarendon Press, Oxford, 174 p.

Goudie A.S. (1985). — Duricrusts and landforms. *In*: Geomorphology and Soils, K.S. Richards, R.R. Arnett, S. Ellis, ed., George Allen, Unwin, London pp. 37–57.

Gruas-Cavagnetto C., Laurain M., Meyer R. (1980a). — Un sol de mangrove fossilisé dans les Lignites du Soissonnais (Yprésien) à Verzenay (Marne). *Geobios.*, 13, 5, pp. 795–801.

Gruas-Cavagnetto C., Laurain M., Meyer R. (1980b). — Paysage végétal et position stratigraphique du sommet des Lignites du Soissonnais dans la Montagne de Reims (Yprésien, Bassin de Paris). *Geobios.*, 13, 6, pp. 947–952.

Guendon J.L. (1981). — Le paléokarst de Coulon (Vaucluse). Thèse, Marseille St. Jérôme, 198 p.

Guendon J.L. (1984). — Les paléokarsts des Alpes Occidentales du Trias à l'Eocène. *Karstologia*, 4, pp. 3–10.

Guendon J.L., Vaudour J. (1981). — Les « tufs » holocènes de Saint-Antonin-sur-Bayon (Bouches-du-Rhône): aspects pétrographiques et signification paléogéographique. *Association Française Karstologie, Mém.* 3, pp. 89–100.

Guendon J.L., Parron C. (1982). — Relations entre la karstification et la bauxitisation dans quelques gisements du Sud-Est de la France. Trav. ERA 282, Institute Géogra. Aix-en-Provence, 11, pp. 69–104.

Guendon J.L., Parron C. (1983). — Bauxites et ocres crétacés du Sud-Est de la France. Trav. Lab. Sci. Terre, St. Jérôme, Marseille, B, 23, 142 p.

Guillet B. (1972). — Relation entre l'histoire de la végétation et de la podzolisation dans les Vosges. Thèse Sci., Nancy, 146 p.

Guillet B., Souchier B. (1979). — Les oxyhydroxydes amorphes et cristallins dans les sols. *In*: Pédologie, 2, Constituants et propriétés du sol, M. Bonneau, B. Souchier, éd.; Masson, Paris, pp. 16–37.

Haguenauer B. (1973). — Contribution de l'analyse séquentielle à la connaissance des formations néogènes du Bassin du Tage au Portugal. Thèse Sci., Nancy, 357 p.

Halitim A., Robert M., Pedro G. (1983). — Etude expérimentale de l'épigénie calcaire des silicates en milieu confiné. Caractérisation des conditions de son développement et des modalités de sa mise en jeu. *Sci. Géol., Mém.*, 71, pp. 63–73.

Halter G., Sheppard S.M.F., Pagel M., Weber F. (1985). — Lithogeochemistry of the alterations correlated with unconformity uranium deposits in the Carswell structure (Saskatchewan, Canada). Symposium SL, Terra Cognita, 5, 2-3, p. 153.

Hanna F.S., Stoops G.J. (1976). — Contribution to the micromorphology of some saline soils of the North Nile delta in Egypt. *Pedologie*, 26, 1, pp. 55-73.

Huang W.Y., Meinschein W.G. (1976). — Sterols as source indicators of organic materials in sediments. *Geochim. Cosmochim. Acta*, 40, pp. 323-330.

Hubert J.F., Reed A.A. (1978). — Red-Bed diagenesis in the East Berlin Formation, Newark Group, Connecticut Valley. *J. Sedim. Petrol.*, 48, 1, pp. 175-184.

Icole M. (1974). — Géochimie des altérations dans les nappes d'alluvions du Piémont occidental nord-pyrénéen. *Sci. Géol., Mém.*, 40, Strasbourg, 200 p.

Iler R.K. (1979). — The Chemistry of Silica. Wiley, New York, 866 p.

James H.L., Dutton C.E., Pettijohn F.J., Wier K.L. (1968). — Geology and ore deposits, Iron River-Crystal Falls district. *Geol. Surv. Prof. Paper*, 570, pp. 72-75.

Jones J.B., Segnit E.R. (1966). — The occurrence and formation of opal at Coober Pedy and Andamooka. *Aust. J. Sci.*, 29, 5, pp. 129-133.

Julia R., Calvet F. (1983). — Descripcion e interpretacion de las texturas y microtexturas de caliches recientes del Camp de Tarragona y Penedes. Libro homenaie J.M. Rios, I.G.M.E., Madrid, pp. 61-96.

Kalliokoski J. (1975). — Chemistry and mineralogy of Precambrian paleosols in Northern Michigan. *Geol. Soc. Am. Bull.*, 86, pp. 371-376.

Kashkaj M.A. (1961). — Gisements d'alunite: leur classification et processus les accompagnant. Izvest. Akad. Nauk, S.S.S.R., Serv. géol., 7, pp. 72-79 (traduction B.R.G.M.).

Kingston D.R., Dish Roon C.P., Williams P.A. (1983). — Global basin classification system. *A.A.P.G. Bull.*, 67, 12, pp. 2175-2193.

Klappa C.F. (1980). — Rhizoliths in terrestrial carbonates: classification, recognition, genesis and significance. *Sedimentology*, 27, pp. 613-629.

Krakenberger. A., Durand M., Even G., Hilly J. (1980). — Etude des altérations au contact socle-couverture dans le Sud-Ouest des Vosges. 105e Cong. Nat. Soc. Savantes, Caen, fasc. II, pp. 207-219.

Kraus M.J., Brown T.M. (1988). — Pedofacies analysis: a new approach to reconstructing ancient fluvial sequences. In: Paleosols and Weathering through Geologic Time, Reinhardt and Sieglo, ed., Special Paper Geol. Soc. Am., 216, pp. 143-152.

Krumbein W.E., Giele C. (1979). — Calcification in a coccoid cyanobacterium associated with the formation of desert stromatolites. *Sedimentology*, 26, pp. 593-604.

Kulbicki G., Vetter P. (1955). — Sur la présence d'argiles bauxitiques dans le Stéphanien de Decazeville. *C.R. Acad. Sci. Paris*, 240, pp. 104-106.

Kulke H. (1974). — Zur Geologie und Mineralogie der Kalkung Gipskrusten Algeriens. *Geol. Rundschau*, 63, 3, pp. 970-998.

Lagny Ph. (1974). — Emersion successives, karstification et sédimentation continentale au Trias moyen dans la région de Sappada. *Sci. Terre*, 19, pp. 193-233.

Lajoinie J.P., Laville P. (1979). — Les formations bauxitiques de la Provence et du Languedoc. Dimensions et distribution des gisements. *Mém. B.R.G.M.*, no. 100, 146 p.

Lajoinie J.P., LAVILLE P. (1980). — Inventaire des formations bauxitiques du Midi de la France. *Annales Mines*, Juillet-Août 1980, pp. 1-12.

Lamouroux M., Loyer J.Y., Bouleau A., Janot C. (1977). — Formes du fer des sols rouges et bruns fersiallitiques. Application de la spectroscopie Mössbauer. *Cah. O.R.S.T.O.M.; sér. Pédol.*, 15, 2, pp. 199-210.

Lang J. (1975). — Un modèle de sédimentation molassique continentale en climat semi-aride: Bassins intramontagneux cénozoïques de l'Afghanistan central. Thèse Sci., Paris 6, 1260 p.

Langford-SMITH T., ed. (1978). — Silcrete in Australia. Dept. Geograph., Univ. New England, Publis., 304 p.

Langford-Smith T., Watts S.H. (1978). — The significance of coexisting siliceous and ferruginous weathering products at select Australian localities. *In*: Silcrete in Australia, T. Langford-Smith ed., Dept. Geograph., Univ. New England, Publis., pp. 143–165.

Laurain M., Meyer R. (1979).— Paléoaltération et paléosol : l'encroûtement calcaire (calcrete) au sommet de la craie. sous les sédiments éocènes de la Montagne de Reims. *C.R. Acad. Sci. Paris*, 289 D, pp. 1211–1214.

Laurain M., Meyer R. (1984). — Paléogéographie thanétienne en Champagne (France). 5e Congr. Europ. Sédim., Marseille, p. 253.

Laurain M., Meyer R. (1986). — Stratigraphie et paléogéographie de la base du Paléocène champenois. *Géol. France, B.R.G.M.*, no. 1, pp. 103–123.

Laville P. (1972). — Géologie, minéralogie, géochimie des formations bauxitiques du Revost-les-Eaux (Var). Contribution à l'étude de la génèse et de l'évolution des bauxites karstiques de Provence. Thèse, Paris VI, 205 p.

Laville P. (1981). — La formation bauxitique provençale (France). Séquence des faciès chimiques et paléomorphologie crétacée. *Chronique Rech. Minière*, 461, pp. 51–68.

Lecolle M. (1967). — Contribution à l'etude de la genèse et de l'évolution des bauxites à mur karstique de Provence. Géologie, minéralogie et sédimentologie des formations bauxitiques de Mazaugues et Pélicon-Merlançon (Var). Thèse, Paris, 135 p.

Lelong F. (1967). — Nature et genèse des produits d'altération de roches cristallines sous climat tropical humide (Guyanne Française). Thèse Sci. Nancy et *Mém. Sci. Terre* 14, 187 p.

Lelong F., Souchier B. (1970). — Bilans d'altération dans la séquence de sols vosgiens, sols bruns acides à podzols, sur granite. *Bull. Serv. Carte Géol. Als. Lorr.*, 23, 3–4, pp. 113–143.

Leprun J.C. (1979). — Les cuirasses ferrugineuses des pays cristallins de l'Afrique de l'Ouest sèche. Genèse, transformation. dégradation. *Sci. Géol., Mém.* 58, 224 p.

Lewan M.D. (1977). — Chemistry and mineralogy of Precambrian paleosols in northern Michigan: discussion. *Geol. Soc. Am. Bull.*, 88, pp. 1375–1376.

Lovering T.S., Engel C. (1967). — Translocation of silica and other elements from rocks into Equisetum and three grasses. *Geol. Survey Prof. Paper*, 594-B, 16 p.

Lucas J., Kalk Y., Gouleau D. (1979). — Aspects minéralogiques et chimiques des sédiments et des sols des mangroves du Sénégal. *Sci. Géol., Mém.* 53, pp. 53–56.

Mabesoone J.M., Lobo H.R. (1978). — Importance of Paleosols for Cenozoic History of Northeastern Brazil. 10th Int. Congress Sedim., Jerusalem, p. 403.

Marzo M., Esteban M., Pomar L. (1974). — Presencia de caliche fósil en el Buntsandstein del valle del Congost (Provincia de Barcelona). *Acta Geol. Hispanica*, IX, 2, pp. 33–36.

Matsumoto R., Iijima A. (1975). — Geochemistry of carbonates in Japanese Paleogene Coal Measures. 9e Congr. Int Sédim. Nice, th. 2, pp. 87–92.

McDonald C.C. (1980). — Mineralogy and geochemistry of a Precambrian regolith in the Athabasca basin. M.Sc. thesis, Saskatchewan University, 151 p.

Mckenzie R.M. (1977). — Manganese oxides and hydroxides. *In*: Minerals in Soil Environments, R.C. Dinauer, ed.; Soil Sci. Soc. Am. Publ., Madison, Wisconsin, pp. 181–193.

Melas P. (1982). — Étude sédimentologique paléogéographique et géochimique du Lias carbonaté du Nord-Lodévois. Thèse Montpellier, et *Mém. C.E.R.G.A.* Montpellier, XIX, 519 p.

Mellor A. (1985). — Soil chronosequences on neoglacial moraine ridges, Jostedalsbreen and Jotunheimen, Southern Norway: a quantitative pedogenic approach. *In*: Geomorphology and Soils, K.S. Richards, R.R. Arnett and S. Ellis, ed., George Allen & Unwin, London, pp. 289–308.

Menillet F. (1984). — Les meulières et les argiles à meulières. Communication à la réunion R.C.P. 706 : « Paléoaltérations et paysages associés » 1 p.

Menillet F. (1993). — Les meulières du Bassin de Paris (France) et les faciès associés. Document 222, BRGM, Orléans, 436 p.

Meyer R. (1973). — La carte géologique au 1/50000 de Rambervillers (Vosges). Présentation générale et commentaires sédimentologiques. Thèse spécialité, Nancy, 150 p.

Meyer R. (1976). — Continental sedimentation, soil genesis and marine transgression in the basal beds of the Cretaceous in the east of the Paris Basin. *Sedimentology*, 23, pp. 235–253.

Meyer R. (1981). — Rôle de la paléoaltération, de la paléopédogenèse et de la diagenèse précoce au cours de l'élaboration des séries continentales. Présentation d'exemples chosis dans quelques formations sédimentaires françaises. Thèse Sci., Nancy, 229 p.

Meyer R. (1984). — Fixation de silice dans les environements continentaux. *Bull. Centres Rech. Explor.-Prod. Elf-Aquitaine,* 8, 1, pp. 195–207.

Meyer R., Guillet B. (1980). — Faciès différenciés d'origine pédologique dans la molasse Oligomiocène d'Aquitaine centrale. *Sci. Géol., Bull.,* 33, 2, pp. 67–80.

Meyer R., Pena Dos Reis R.B. (1985). — Paleosols and alunite silcretes in continental Cenozoic of Western Portugal. *Jour. Sedim. Petrol.,* 55, 1, pp. 76–85.

Michel P. (1978). — Cuirasses bauxitiques et ferrugineuses d'Afrique occidentale. Apercu chronologique. *In*: Géomorphologie des reliefs cuirassés dans les pays tropicaux chauds et humides. *Travaux et Documents de Géographie Tropicale* C.E.G.E.T. Bordeaux no. 33.

Milliken K.L. (1979). — The silicified evaporite syndrome. Two aspects of silicification history of former evaporite nodules from Southern Kentucky and Northern Tennessee. *Jour. Sedim. Petrol.,* 49, 1, pp. 245–256.

Millot G. (1949). — Relations entre la constitution et la genèse des roches sédimentaires argileuses. Thèse Sci., Nancy, 352 p.

Millot G. (1964). — Géologie des argiles. Masson, Paris, 499 p.

Millot G. (1977). — Géochimie de la surface et formes du relief. Présentation. *Sci. Géol., Bull.,* 30–4, pp. 229–233.

Millot G. (1982). — Weathering sequences. « Climatic » planations. Leveled surfaces and paleosurfaces. *In*: International Clay Conference 1981, Olphen H. van, Veniale F. ed., Elsevier, Amsterdam, pp. 585–593.

Millot G., Nahon D., Paquet H., Ruellan A., Tardy Y. (1977). — L'épigénie calcaire des roches silicatées dans les encroûtements carbonatés en pays subaride, Antiatlas, Maroc. *Sci. Géol., Bull.,* 30, 4, pp. 129–152.

Mitchell B.D. (1975). — Oxides and hydrous oxides of silicon. *In*: Soil Components, vol. 2, Inorganic components, J.E. Gieseking, ed., Springer-Verlag, Berlin, pp. 395–432.

Mount J.F., Cohen A.S. (1984). — Petrology and geochemistry of rhizoliths from Plio-Pleistocene fluvial and marginal lacustrine deposits, East Lake Turkana, Kenya. *Jour. Sedim Petrol.,* 54, 1, pp. 263–275.

Nahon D. (1976). — Cuirasses ferrugineuses et encroûtements calcaires au Sénégal occidental et en Mauritanie. Systèmes évolutifs: géochimie, structure, relais et coexistence. Thèse Sci., Marseille et *Sci. Géol., Mém.* 44, 232 p.

Nahon D., Lappartient J.R. (1977). — Time factor and geochemistry in iron crusts genesis. *Catena,* 4, pp. 249–254.

Nahon D., Millot G. (1977). — Enfoncement géochimique des cuirasses ferrugineuses par épigénie du manteau d'altération des roches-mères gréseuses. Influence sur le paysage. *Sci. Géol., Bull.,* 30, 4, pp. 275–282.

Nahon D., Ducloux J., Butel P., Augas C., Paquet, H. (1980). — Néoformation d'aragonite, première étape d'une suite minéralogique évolutive dans les encroûtements calcaires. *C.R. Acad. Sci. Paris,* 291 D, pp. 725–727.

Netterberg F. (1974). — Calcretes and silcretes at Sambio, Okavangoland, South West Africa. *S. Afr. Archaeol. Bull.,* 29, pp. 83–88.

Netterberg F. (1978). — Rates of calcrete formation. 10th Int. Congress Sedim., Jerusalem, p. 465.

Nicolas J., Bildgen P. (1979). — Relations between the location of the Karst bauxites in the northern hemisphere, the global tectonics and the climatic variations during geological time. *Paleogeogr., Paleoclim., Paleoecol.,* 28, pp. 205–239.

Ortlam D. (1971). — Paleosols and their significance in stratigraphy and applied geology in the Permian and Triassic of Southern Germany. *In*: Paleopedology, D. Yaalon, ed., Israel Univ. Press, Jerusalem, pp. 321–327.

Pagel M. (1983). — Les altérations *In*: Les Gisements d'Uranium Liés Spatialement aux Discordances, M. Pagel, ed., *Mém. C.R.E.G.U.,* 1, Nancy, pp. 289–302.

Paquet H. (1969). — Évolution géochimique des minéraux argileux dans les altérations et les sols des climats méditerranéens et tropicaux à saisons constrastées. Thèse Sci., Strasbourg, et *Mém. Serv. Carte géol. Als. Lorr.,* 30, 210 p.

Paquet H. (1983). — Stability, instability and significance of attapulgite in the calcretes of Mediterranean and tropical areas with marked dry season. *Sci. Géol., Mém.,* 72, pp. 131–140.

Paquet H., Ruellan A. (1993). — Epigénie et encroûtements calcaires (calcrètes). Colloque G. Millot, Paquet and Clauer, eds. Mémoire Acad. Sci., Paris, pp. 19–39.

Parron C., Nahon D., Tardy Y. (1980). — Cuirassements ferrugineux et barytiques par altération latéritique des grès hettangiens de la région de Chaillac (Indre). *Mém. B.R.G.M.,* 104, pp. 385–405.

Parron C., Nahon D., Fritz B., Paquet H., Millot G. (1976). — Désilicification et quartzification par altération des grés albiens du Gard. Modèles géochimiques de la genèse des dalles quartzitiques et silcrètes. *Sci. Géol., Bull.,* 29, 4, pp. 273–284.

Parry W.T., Reeves C.C. (1968). — Clay mineralogy of pluvial lake sediments, southern High Plains, Texas. *Jour. Sedim. Petrol.,* 38, 2, pp. 516–529.

Pedro G. (1968). — Distribution des principaux types d'altération chimique à la surface du globe. *Rev. Géogr. Phys. Géol. Dynam.,* (2), X, 5, pp. 457–470.

Pedro G. (1979). — Les conditions de formation des constituants secondaires. *In*: Pédologie 2, Constituants et Propriétés du Sol, M. Bonneau, B. Souchier, éd., Masson, Paris, pp. 58–72.

Pedro G. (1993). — Argiles des altérations et des sols. Colloque G. Millot, Paquet and Clauer, eds., Mémoire Acad. Sci., Paris, pp. 1–17.

Pedro G., Delmas A.B., Seddoh F.K. (1975). — Sur la nécessité et l'importance d'une distinction fondamentale entre type et degré d'altération. Application au problème de la définition de la ferrallitisation. *C.R. Acad. Sci. Paris,* 280 D, pp. 825–828.

Pena Dos Reis R.B. (1983). — A sedimentologia de depositos continentais. Dois exemplos do Cretácico superior-Miocénico de Portugal. Thèse Sci., Coimbra, 404 p.

Pena Dos Reis R.B., Meyer R. (1982). — Sédimentation continentale du Crétacé terminal au Miocène dans le Bassin de Coimbra-Leiria (Portugal). Actions tectoniques et climatiques (silicifications). *C.R. Acad. Sci. Paris,* II, 294, pp. 741–744.

Perel'man A.I. (1967). — Geochemistry of Epigenesis. Plenum Press, New York, 266 p.

Perin G. (1974). — Contribution à la connaissance minéralogique des formations bauxitiques de Provence. Thèse Sci., Marseille, 212 p.

Perinet G., Taieb M., Tiercelin J.J. (1980). — Présence de natrojarosite dans les sédiments lacustres d'Hadar (Éthiopie). 8e Réunion an. Sci. Terre, Marseille, p. 276.

Pettijohn F.J. (1957). — Sedimentary Rocks, 2nd edition, Harper, New York, 718 p.

Plaziat J.C. (1970). — Huîtres de mangrove et peuplements littoraux de l'Eocène inférieur des Corbières. *Geobios.,* 3, pp. 7–27.

Plaziat J.C. (1971). — Racines ou terriers? Critères de distinction à partir de quelques exemples du Tertiaire continental et littoral du Bassin de Paris et du midi de la France. Conséquences paléogéographiques. *Bull. Soc. Géol. France,* (7), XIII, pp. 195–203.

Plaziat J.C. (1975). — Les mangroves anciennes. Discussion de leurs critères de reconnaissance et de leurs significations paléoclimatologiques. 9e Congr. Int. Sédim., Nice, th, 1, pp. 153–159.

Plaziat J.C. (1984). — Le biotope palustre: méthode d'identification paléoécologique. *Geobios. Mém.* spécial no. 8, pp. 313–320.

Plaziat J.C., Freytet P. (1978). — Le pseudo-microkarst pédologique: un aspect particulier des paléopédogenèses dévelopées sur les dépôts calcaires lacustres dans le Tertiaire du Languedoc. *C.R. Acad. Sci. Paris,* 286 D, pp. 1661–1664.

Plaziat J.C., Koeniger J.C., Baltzer F. (1983). — Des mangroves actuelles aux mangroves anciennes. *Bull. Soc. Géol. France,* 7, t. XXV, no. 4, pp. 499–504.

Plet-Lajoux C., Monnier G., Pedro G. (1971). — Étude expérimentale sur la genèse et la mise en place des encroûtements gypseux. *C.R. Acad. Sci. Paris,* 272, pp. 3017–3020.

Pomerol C. (1964). — Découverte de paléosols de type podzol au sommet de l'Auversien (Bartonien inférieur) de Moisselles (Seine et Oise). *C.R. Acad. Sci. Paris,* 258, pp. 974–976.

Pons L. (1972). — Outline of the genesis, characteristics, classification and improvement of acid sulphate soils. *In*: Acid Sulphate Soils Symp., publication 18, Inter. Inst. for Land Reclam. and Improv., Wageningen, pp. 3–27.

Pouget M. (1968). — Contribution à l'étude des croûtes et encroûtements gypseux de nappe dans le Sud Tunisien. *Cah. O.R.S.T.O.M., sér. Pédol.*, VI, pp. 309–365.

Reeves C.C. (1976). — Caliche. Origin, classification, morphology and uses. Estacado Books, Lubbock, Texas, 233 p.

Reinhardt J., Sieglo W.R. (1988).— Paleosols and weathering through geologic time: Principles and applications. Special Paper Geol. Society America, 216, 285 p.

Retallack G. (1981a). — Comment on «Reinterpretation of the depositional environment of the Yellowstone fossil forests». *Geology*, 9, pp. 52–53.

Retallack G. (1981b). — Fossil soils: indicators of ancient terrestrial environments. *In:* Paleobotany, Paleoecology and Evolution; K.J. Niklas, ed.; vol. 1, Praeger, New York, pp. 55–102.

Retallack G. (1983). — Late Eocene and Oligocene paleosols from Badlands National Park, South Dakota. *Geol. Soc. America, Spec. Paper* 193, 82 p.

Retallack G. (1984). — Completeness of the rock and fossil record: some estimates using fossil soils. *Paleobiology*, 10, 1, pp. 59–78.

Retallack G. (1986). — Precambrian paleopedology. Special Issue, *Precambrian Research*, 32, 2–3, 259 p.

Retallack G.J. (1989). — Paleosols and their relevance to Precambrian atmospheric composition: comment. *J. Geology*, 97, pp. 763–764.

Retallack G.J. (1991). — Miocene paleosols and ape habitats of Pakistan and Kenya. Oxford University Press, New York, 346 p.

Retallack G.J., Wright V.P. (1990). — Micromorphology of lithified paleosols. *In*: Soil Micromorphology: A Basic and Applied Science, L.A. Douglas, ed. Elsevier, Amsterdam, pp. 641–652.

Retallack G., Grandstaff D., Kimberley M. (1984). — The promise and problems of Precambrian paleosols. *Episodes*, 7, 2, p. 812.

Richter D.K., Füchtbauer H. (1978). — Ferroan calcite replacement indicates former magnesian calcite skeletons. *Sedimentology*, 25, pp. 843–860.

Robinson D., Wright V.P. (1987). — Ordered illite-smectite and kaolinite-smectite: pedogenic minerals in a Lower Carboniferous paleosol sequence, South Wales. *Clay Minerals*, 22, pp. 109–118.

Rocha F., Gomez C. (1992). — Clay mineralogy of paleosurfaces in the Tertiary and Quaternary of the Aveiro sedimentary basin. *In*: Mineralogical and Geochemical Records of Paleoweathering, IGCP 317, Schmitt and Gall, eds., ENSMP Mém. Sci. Terre, 18, pp. 31–38.

Röper H.P., Rothe P. (1975). — Petrology of fossil duricrust: the «Karneoldolomit Horizont», Permian, SW-Germany. 9th. Int. Sedim. Congress, Nice, thème 2, pp. 113–118.

Ross G.M., Chiarenzelli J.R. (1985). — Paleoclimatic significance of widespread Proterozoic silcretes in the Bear and Churchill provinces of the Northwestern Canadian Shield. *Jour. Sedim. Petrol.*, 55, 2, pp. 196–204.

Routhier P. (1963). — Les gisements métallifères (Tome 1), Masson, Paris, 867 p.

Roy S. (1981). — Manganese deposits. Academic Press, London; 365 p.

Rozanov A.N. (1951). — The serozems of Central Asia. Israel Progr. Sci. Translation, Jérusalem, 472 p.

Ruellan A. (1970). — Les sols à profil calcaire différencié des plaines de la Basse Moulouya (Maroc oriental). Thèse Sci., Strasbourg et *Mém. O.R.S.T.O.M.*, 54, 302 p.

Sassi S., Triat J.M., Truc G. (1984). — Les encroûtements éocènes (calcrètes et dolocrètes) à Bulimes de Tunisie; genése et importance paléogéographique de leur découverte. 5e Congr. Europ. Sédim., Marseille, p. 397.

Schau M., Henderson J.B. (1983). — Archean chemical weathering at three localities on the Canadian shield. *Precambrian Research*, 20, pp. 189–224.

Schmitt J.M. (1983). — Albitization in relation to the formation of uranium deposits in the Rouergue area (Massif Central, France). *Sci. Géol. Mém.* 73, pp. 185–194.

Schneebeli W. (1976). — Untersuchungen von Gletscherschwankungen im Val de Bagnes. *Die Alpen*, Lucerne, 1976, 3–4, pp. 5–57.

REFERENCES

Schroeder D. (1973). — Zur Stellung der Gleye and Pseudogleye in verschiedenen Klassifizierungssystem. *In*: Pseudogley and Gley, E. Schlichting, V. Schwertmann, ed., Verlag Chemie, Weinheim, pp. 413–419.

Schwertmann U., Taylor R.M. (1977). — Iron oxides. *In*: Minerals in Soil Environments, R.C. Dinauer, ed.; Soil Sci. Soc. Am. Publ., Madison, Wisconsin, pp. 145–180.

Scurfield G., Segnit E.R. (1984). — Petrification of wood by silica minerals. *Sedim. Geology*, 39, pp. 149–167.

Sherman G.D., Schultz F., Alway F.J. (1962). — Dolomitization in soils of the Red River Valley, Minnesota. *Soil Science*, 94, pp. 304–313.

Siedlecka A. (1976). — Silicified Precambrian evaporite nodules from Northern Norway: a preliminary report. *Sedim. Geology*, 16, pp. 161–175.

Simon-Coicon R., Thiry M., Schmitt J.M., Legendre S., Astruc J.G. (1995). — From inland paleosurfaces towards sedimentary basins. The example of Southern French Massif Central. 16th IAS Regional Meeting, Chambery, France, Field Trip Book 23, 64 p.

Sloss L.L., Speed R.C. (1974). — Relationships of cratonic and continental-margin tectonic episodes. *In*: Tectonics and Sedimentation, Dickinson, ed., Soc. Econ. Paleontol. Mineralo. Spec. Publ. 22, Tulsa, pp. 98–119.

Smale D. (1973). — Silcretes and associated silica diagenesis in Southern Africa and Australia. *Jour. Sedim. Petrol.*, 43, 4, pp. 1077–1089.

Soil Survey Staff (1975). — Soil Taxonomy. U.S.D.A. Handbook no. 436, 754 p.

Souchier B. (1971). — Évolution des sols sur roches cristallines à l'étage montagnard (Vosges). Thèse Sci. Nancy et *Mém. Serv. Carte Géol. Als. Lorr.* 33, Strasboug, 134 p.

Souchier B. (1981). — Résumé de la présentation des thèmes du colloque. *In*: Migrations Organo-minérales dans les Sols Tempérés. Colloques Internat. C.N.R.S. no. 303, pp. 19–20.

Souchier B., Lelong F. (1970). — Détermination quantitative par voie chimique des constituants minéralogiques des sols tempérés. *Bull. Serv. Carte Géol. Als. Lorr.*, 23, 3–4, pp. 95–112.

Steinberg M. (1970). — Contribution à la sédimentation et de la géochimie à l'étude des formations continentales azoïques. Faciès sidérolithique du détroit poitevin. *C.T.H.S., Mém. Sci.* 3, 175 p.

Stoops G.J., Eswaran H., Abtahi A. (1977). — Scanning electron microscopy of authigenic sulfate minerals in soils. *In*: 5th Int. Working Meeting on Soil Micromorphology, Delgado, ed., Granada (Spain), pp. 1093–1113.

Straaten L. van (1978). — Dendrites. *Jour. Geol. Society*, 135, pp. 137–151.

Summerfield M.A. (1983a). — La silcrète en Australie, en Afrique australe et en Grande-Bretagne: une revue de la littérature anglaise. *Rev. Géol. Dynam. Géog. Phys.*, 24, 5, pp. 397–410.

Summerfield M.A. (1983b). — Petrography and diagenesis of silcrete from the Kalahari Basin and Cape coastal zone, Southern Africa. *Jour. Sedim. Petrol.* 53, 3, pp. 895–909.

Summerfield M.A. (1983c). — Silcrete as a palaeoclimatic indicator: evidence from southern Africa. *Palaeogeog., Palaeoclim., Palaeoecol.*, 41, pp. 65–79.

Surdam R.C., Eugster H.P., Mariner R.H. (1972). — Magaditype chert in Jurassic and Eocene to Pleistocene rocks, Wyoming. *Geol. Soc. Am. Bull.*, 83, pp. 2261–2266.

Swett K. (1974). — Calcrete crusts in an Arctic Permafrost environment. *Am. Jour. Science*, 274, pp. 1059–1063.

Tardy Y. (1969). — Géochimie des altérations. Étude des arènes et des eaux de quelques massifs cristallins d'Europe et d'Afrique. Thèse Sci., Strasbourg et *Mém. Serv. Carte Géol. Als. Lorr.* 31, Strasbourg, 199 p.

Tardy Y. (1993). — Climats, paléoclimats et biogéodynamique du paysage tropical. Colloque G. Millot, Paquet and Clauer, eds., Mémoire Acad. Sci. Paris, pp. 141–175.

Teruggi M.E., Andreis R.R. (1971). — Micromorphological recognition of paleosolic features in sediments and sedimentary rocks. *In*: Paleopedology, D. Yaalon, ed., Israel Univ. Press, Jerusalem, pp. 161–172.

Thiry M. (1981). — Sédimentation continentale et altérations associées: calcitisations, ferruginisations et silicifications. Les argiles plastiques du Sparnacien du Bassin de Paris. *Sci. Géol., Mém.*, 64, 173 p.

Thiry M., Turland M. (1984). — Paléotoposéquences de sols ferrugineux et de cuirassements siliceux dans le Sidérolithique du Nord du Massif Central (Bassin de Montluçon-Domerat). Communication Réunion R.C.P. 706: Paléoaltérations et Paysages Associés, 2 p.

Thiry M., Milnes A.R. (1991). — Pedogenic and ground water silcretes at Stuart Creek Opal Fields, South Australia. *J. Sedim. Petrology*, 61, pp. 111–127.

Thiry M., Panziera J.P., Schmitt J.M. (1984). — Silicification et désilicification des Grès et des Sables de Fontainebleau. Évolutions morphologiques des grès dans les sables et à l'affleurement. *Bull. Inf. Géol. Bass. Paris*, 21, 2, pp. 23–32.

Thiry M., Schmitt J.M., Trauth N., Cojean R., Turland M. (1983). — Formations rouges «sidérolithiques» et silicifications sur la bordure nord du Massif Central. *Revue Géol. Dynam. Géogra. Phys.*, 24, 5, pp. 381–395.

Tissot B.P., Welte D.H. (1978). — Petroleum Formation and Occurrence. Springer-Verlag, Berlin, 538 p.

Toulemont M. (1984). — Le Karst gypseux du Lutétien supérieur de la région parisienne. Caractéristiques et impact sur le milieu urbain. *Rev. Géol. Dynam. Géogr. Phys.*, 25, 3, pp. 213–228.

Trescases J.J. (1975). — L'évolution géochimique supergène des roches ultrabasiques en zone tropicale. Formation des gisements nickelifères de Nouvelle-Calédonie. *Mém. O.R.S.T.O.M.*, 78, Paris, 254 p.

Triat J.M. (1979). — Paléoaltérations dans le Crétacé supérieur de Provence rhodanienne. Thèse Sci., Marseille, et *Sci. Géol., Mém.* 68, 202 p.

Troy J.P. (1979). — Pédogenèse sur roches charnockitiques en région tropicale humide de montagne, dans le sud de l'Inde. Thèse Sci., Nancy, 366 p.

Truc G. (1975). — Sols à profil calcaire différencié et pellicules rubanées dans le Paléogène du Sud-Est de la France. In: Colloque sur les Croûtes Calcaires et Leur Répartition Régionale, T. Vogt, ed., Strasbourg, pp. 108–113.

Turner P. (1980). — Continental Red Beds. Elsevier, Amsterdam, 561 p.

Twidale C.R. (1983). — Australian laterites and silcretes: ages and significance. *Rev. Géol. Dynam. Géogr. Phys.*, 24, 1, pp. 35–45.

Urbani F. (1978). — Les Karsts gréseux du Venezuela. *Spelunca*, 1, pp. 24–28.

Valeton I. (1972). — Bauxites. Elsevier, Amsterdam, 226 p.

Valleron M.M. (1981). — Les faciès calcaires du Lutétien à Planorbis pseudoammonius du Bas-Languedoc. Thèse, Strasbourg, 122 p.

Van Breemen N. (1972). — Soil forming processes in acid sulphate soils. In: Acid Sulphate Soils Symp., Publication 8, Inter. Inst. for Land Reclam. and Improv., Wageningen, pp. 66–129.

Van Wambeke A. (1973). — Hydromorphie et formation de plinthite dans les sols de plaines herbeuses de Colombie. In: Pseudogley and Gley, E. Schlichting, V. Schwertmann, ed., Verlag Chemie, Weinheim, pp. 357–362.

Vatan A. (1947). — La sédimentation continentale tertiaire dans le Bassin de Paris méridional. Thèse Sci. Éd. Toulous Ingénieur, 215 p.

Veneman P.L.M., Vepraskas M.J., Bouma J. (1976). The physical significance of soil mottling in a Wisconsin toposequence. *Geoderma*, 15, pp. 103–118.

Verrecchia E.P. (1990). — Litho-diagenetic implications of the calcium oxalate-carbonate biochemical cycle in semiarid calcretes, Nazareth, Isarel. *Geomicrobiology J.*, 8, pp. 87–99.

Verrecchia E.P. (1991). — Stromatolitic origin for desert laminar limecrusts. *Naturwissenschaften* 78, pp. 505–507.

Vervier P. (1990). — Hydrochemical characterization of the water dynamics of karstic system. *J. Hydrol.*, 121, pp. 103–117.

Vetter P., Hery B., Laversanne J. (1975). — Sédimentation continentale dans les bassins houillers et permiens du Sud du Massif Central. Guide Excursion no. 22, 9e Congr. Int. Sédim., Nice, 35 p.

Vieillefon J. (1974). — Contribution à l'étude de la pédogenèse dans le domaine fluvio-marin en climat tropical d'Afrique de l'Ouest. Thèse Sci., Paris 5, 346 p.

REFERENCES

Wackermann J.M. (1975). — L'altération des massifs cristallins basiques en zone tropicale semi-aride. Etude minéralogique et géochimique des arènes du Sénégal oriental. Conséquences pour la cartographie et la prospection. Thèse Sci., Strasbourg, 373 p.

Wada K. (1977). — Allophane and imogolite *In*: Minerals in Soil Environments, R.C. Dinauer, ed.; Soil Sci. Soc. Am. Publ., Madison, Wisconsin, pp. 603–638.

Wakatsuki T., Furukawa H., Kyuma K. (1977). — Geochemical study of the redistribution of elements in soil. I. Evaluation of degree of weathering of transported soil materials by distribution of major elements among the particle size fractions and soil extract. *Geochim. Cosmochim. Acta*, 41, pp. 891–902.

Walker T.R., Waugh B., Crone A.J. (1978). — Diagenesis in first-cycle desert alluvium of Cenozoic age, southwestern United States and northwestern Mexico. *Geol. Soc. Am. Bull.*, 89, pp. 19–32.

Watts N.L. (1977). — A comparative study of some Quaternary, Permo-Triassic and Siluro-Devonian calcretes. Ph.D., Reading 265 p.

Watts N.L. (1978). — Displacive calcite: Evidence from recent and ancient calcretes. *Geology*, 6 pp. 699–703.

Weaver C.E., Pollard L.D. (1973). — The Chemistry of Clay Minerals. Elsevier, Amsterdam, 213 p.

Wetzel A. (1985). — Asymmetry of Zoophycos burrows as a way-up criterion. A reconsideration. *Sedimentology*, 32, pp. 749–751.

Wilding L.P., Smeck N.E., Drees L.R. (1977). — Silica in soils: quartz, cristobalite, tridymite and opal. *In*: Minerals in Soil Environments, R.C. Dinauer, ed.; Soil Sci. Soc. Am. Publ., Madison, Wisconsin, pp. 471–552.

Williams C., Yaalon D.H. (1977). — An experimental investigation of reddening in dune sand. *Geoderma*, 17, pp. 181–191.

Williams G.E. (1968). — Torridonian weathering and its bearing on Torridonian palaeoclimate and source. *Scott. J. Geol.* 4, pp. 164–184.

Wilson M.D., Pittman E.D. (1977). Authigenic clays in sandstones: recognition and influence on reservoir properties and paleoenvironmental analysis. *Jour. Sedim. Petrol.*, 47, 1, pp. 2–31.

Wright V.P. (1982a). — The recognition and interpretation of paleokarsts: two examples from the Lower Carboniferous of South Wales. *Jour. Sedim. Petrol.*, 52, 1, pp. 83–94.

Wright V.P. (1982b). — Calcrete paleosols from the Lower Carboniferous Llanelly Formation, South Wales. *Sedim. Geology*, 33, pp. 1–33.

Wright V.P. (1983). — A rendzina from the Lower Carboniferous of South Wales. *Sedimentology*, 30, 2, pp. 159–180.

Wright P., ed. (1986). — Paleosols: Their Recognition and Interpretation. Blackwell, Oxford, 312 pp.

Wyns R. (1991). — Structural evolution of the Armorican basement during the Cenozoic deduced from analysis of continental paleosurfaces and associated deposits. *Géologie France*, 3–1991, pp. 11–42.

Yaalon D.H. (1970). — Parallel stone cracking, a weathering process on desert surfaces. *Geol. Inst. Techn. Econ. Bull.*, Bucharest, C, 18, pp. 107–111.

Yaalon D.H. (1971). — Soil-forming processes in time and space. *In*: Paleopedology, Yaalon, ed., Israel Univ. Press, Jerusalem, pp. 29–39.

Yaalon D.H. (1978). — «Geoderma», continental sedimentation, calcrete, desert loess and paleosols, sand dunes and eolianites. Guide Book Y5, 10th Int. Congress Sedim, Jerusalem, pp. 195–238.

Yaalon D.H., Singer S. (1974). — Vertical variation in strength and porosity of calcrete (Nari) on chalk, Shefela, Israel, and interpretation of its origin. *Journ. Sedim. Petrol.*, 44, 4, pp. 1016–1023.

Yuretich R.F. (1984) — Yellowstone fossil forests: new evidence for burial in place. *Geology*, 12, pp. 159–162.

Zeegers H., Leprun J.C. (1979). — Évolution des concepts en altérologie tropicale et conséquences potentielles pour la prospection géochimique en Afrique occidentale soudano-sahélienne. *Bull. B.R.G.M* (2), 2–3, pp. 229–239.

INDEX

Accretion 25, 32, 33, 39, 90
Accumulation (*see* Argillaceous, Chemical elements, Clay etc.)
Acidolysis 89–90, 118, 119, 123, 128
Aeolian 8, 20, 32, 33, 34, 59, 93
Agglomeratic fabric 129, 130
Albite (albitisation) 81–84, 85, 106, 128
Alcrete 20, 114, 119
Allitisation 89, 90, 127
Allochthon (alterite) 3, 69, 70, 73
Allochthony 3
Allophane 77, 79
Alteration (*also see* paleoprofile, profile, weathering) 2, 4, 62, 63, 64, 66, 67, 71, 76, 77, 84, 101, 108, 110, 126, 128
Alterite 2, 4, 61, 62, 119–123
— superimposition of 4, 71–72, 94
Alunite 32, 36–39, 45, 46, 47, 69, 75, 103, 113
Analcime 103
Andosol (andosolisation) 41, 77, 78, 79, 114
Anhydrite 16, 33, 34, 35, 39, 50, 103
Argillaceous accumulation 11–15, 23, 29, 43, 44, 45, 47, 48, 50, 61, 69, 81, 95, 96, 105
Argillan 10–13, 15, 110, 112, 131, 132
Authigenesis (*also see* Neoformation) 14, 15, 30, 41, 44, 46, 61, 81, 101, 104, 105, 118
Autochthon (alterite) 3, 62, 70, 73
Authocthony 3, 62

Barite 33, 36–39, 103
Basalt 4, 16, 22, 31, 67, 68, 77, 78
Bauxite 16, 17, 39, 65, 67, 68, 69, 70–75, 88–89, 97, 99, 120
Biorhexistasy 1
Biotite 10, 80, 83, 85, 106
Bisiallitisation 89, 90, 127
Boehmite 67, 69, 71, 72
Boxworks 33, 34, 39, 45, 47, 49, 91, 98
Burial 15, 20, 29, 30, 34, 37, 49–51, 59, 60, 80, 81, 84, 94, 99, 101, 102, 113, 116, 117
Burrows 7–9, 45, 109, 132

Calcedonite 40, 119
Calcitan 132
Calcite 16, 22, 23, 24, 31, 34, 43, 50, 91, 103, 132
Calcrete 16, 19, 20–31, 53–54, 55, 84, 91, 93–94, 98, 102, 103, 114, 119, 120, 124
Caliche 20
Carapace 61, 62, 64, 67
Carbonates 6, 10, 11, 13, 15–32, 33, 34, 42, 43, 47, 49, 50, 53, 55, 56, 65, 69, 71, 73, 76, 95, 99, 101, 104, 107, 112, 113, 116, 120, 127, 128
— accumulation 26
Cartography 107, 115
Cavities (*see* Porosity)
Celestite 33, 103
Cementation 20, 21, 25, 41, 44, 45, 51, 54, 56, 90, 95, 98, 102, 104
Chalk 23, 24, 51, 52, 94, 102
Chemical elements 123
— accumulation 5, 48, 60, 73, 90, 91, 101
— migration 5, 91
Chernozem 120, 124, 127
Chlorite (chloritisation) 10, 14, 81, 85
Classification (paleosols) 111–112
Clay
— accumulation 5, 9, 11, 13–15, 27, 28, 76, 89, 96, 100, 104, 112, 113, 121, 128
— migration 5, 9, 11
— redistribution 5, 10
Climate 4, 5, 10, 15, 18, 19, 20, 21, 25, 31, 32, 39, 40, 45, 55, 56, 57, 59, 60, 61, 62, 87, 89–90, 91, 93, 94, 97, 99, 113–114, 120, 122–125, 128
— arid 15, 19, 20, 22, 25, 34, 39, 62, 90, 114, 122, 123, 125, 127, 128
— boreal 114, 128
— desert 20, 31, 59, 89, 92, 102
— equatorial 4, 56, 89, 114
— Mediterranean 20, 21, 31, 56, 89, 113, 114, 128
— temperate 10, 15, 57, 67, 89, 98, 113, 114, 118, 120, 127, 128
— tropical 10, 56, 59, 89, 91, 113, 114, 128
— warm and humid 45, 62, 66, 67, 73, 75, 76, 77, 89, 90, 94, 98, 103, 118, 122, 123, 125, 127, 128

Climatic zonation 89–90, 117–119
Climax 117–119, 120
Colour (soils) 11, 12, 13, 14, 15, 21, 23, 28, 31, 37, 42, 43, 44, 45, 53, 55, 56, 57, 59, 61, 63, 64, 65, 66, 67, 75, 76, 77, 81, 82, 83, 85, 98, 102, 103, 109, 113, 114, 127–128
Compaction 7, 8, 18, 33, 102, 112, 113, 114, 116
Creep 7
Crust 2, 18, 19, 20–31, 32, 39, 41, 42, 43, 45, 49, 50, 53, 55, 76, 91, 96, 98, 100, 101, 102, 114, 118, 119, 128
Cuirasse (*also see* Duricrust) 21, 39, 56, 58–67, 122, 123
Cutan 116, 129, 131–132
— stress cutan 15, 112, 131, 132
Cyclothem 37, 91–93

Dating (*see* Soil age)
Decarbonation 120
Desilicification 42, 63
Diagenesis 5, 7, 9, 10, 15, 16, 20, 25, 29, 30, 31, 37, 39, 42, 49–51, 54, 57, 59, 66, 75, 77, 80, 81, 84, 85, 98, 100, 101–102, 105, 107, 110, 116, 126
Diaspore 67, 69
Diastem 3, 7, 13, 14, 23, 29, 36, 44, 47, 85, 91, 92, 97
Differentiation, pedological 7, 9, 32, 34, 36, 76, 77, 115, 120, 121, 124, 125
Discordance 1, 81, 82, 84, 87, 88, 90, 94–96, 98, 99, 104, 108, 120, 126
Dolocrete 20–31, 28, 29, 42, 49, 53–54, 99, 103, 114
Dolomite (dolomitisation) 16, 26, 28–31, 33, 43, 49, 73, 96, 97, 103, 128
Downstream zone 10, 28, 30, 48, 88, 90, 101, 104
Drainage 10, 45, 48, 56, 57, 60, 61, 66, 69, 73, 118, 119, 120, 123, 125, 128
Duricrust (*also see* Cuirasse) 98, 99, 120
Duripan 113, 114

Earthworms 7–8
Eh 75, 118, 123, 124
Eluviation 112, 121, 129
Emergence 4, 6, 9, 16, 18, 23, 25, 34, 39, 59, 63, 66, 69, 74, 77, 84, 87, 89, 90, 94–96, 110
Epigenesis 9, 22, 29, 34, 36, 39, 47, 51, 53, 55, 59, 60
Erosion 1, 4, 13, 18, 19, 25, 48, 60, 61, 62, 71, 73, 76, 80, 81, 84, 87, 88, 90, 94–96, 97, 98, 99, 100, 108, 110, 123
Evaporite 14, 32, 54, 55, 56, 91, 93, 96, 103

Evolution, degree (duration) of 1, 13, 14, 17, 18, 23, 26, 29, 32, 34, 39, 44, 47, 53, 55, 57, 58, 60, 69, 71, 77, 78, 79, 84, 86, 88, 89, 90, 91, 94, 96–99, 105, 106, 107, 110, 116, 117–118, 119–122, 123, 125, 128

Ferrallite 69, 89, 90, 114, 120, 122, 127
Ferrallitisation 62, 67, 69, 84, 86, 89, 90
Ferran 132
Ferri-argillan 10–13, 15, 132
Ferricrete 20, 67, 98, 114, 119, 124
Ferruginisation 44, 56–67, 96, 98, 120, 123
Flagstone 21, 65
Flint clay 23, 51, 52, 69, 70
Fossil forests 5, 9, 118
Fragipan 112

Geochemical barriers 100–101, 104
Geochemistry (geochemical context) 34, 61, 68, 75, 78, 86, 87, 89–91, 94, 98, 101–102, 105–107, 109, 110, 115, 116, 121, 123
Geomorphology 59, 100, 110
Gibbsan 132
Gibbsite 60, 67, 68, 71, 72, 78, 89, 90, 103, 112, 127, 132
Gilgai 50, 69
Glaebule 132
Gley 84, 111, 120, 123, 124, 127
Gloss (glossic soil) 51, 52, 57, 59, 128
Goethite 34, 44, 53, 56, 57, 58, 59, 60, 63, 65, 66, 67, 68, 86, 98, 102, 103, 119
Granite 4, 31, 63, 81, 120
Granular (fabric) 23, 34, 52, 53, 109, 129, 130
Gypcrete 20, 32, 33, 103, 114, 119, 124
Gypsum 16, 32–39, 50, 54, 69, 90, 103, 114

Haematite 42, 51, 56, 58, 59, 60, 63, 65, 66, 67, 68, 85, 86, 102, 103, 119
Halloysite 13, 14, 78, 103
Horizons 8, 9, 11, 12, 13, 15, 21, 25, 33, 34, 36, 39, 41, 55, 56, 59, 60, 64, 65, 77, 78, 86, 100, 102, 109, 111–115, 117, 118, 119, 120–122, 123, 124, 125, 127, 128
— albic 112, 114, 128
— alumino-phosphatic 114
— argillic 53, 78, 112, 113, 114, 127, 128
— calcic 112, 113, 114, 128
— cambic 112, 113, 114, 120, 128
— diagnostic 109, 111–115, 118
— eluvial 63, 111
— gypsic 103, 112, 114
— histic 8, 100, 112, 114
— illuvial 63, 111

— mollic 112, 114
— natric 112, 114, 128
— oxic 33, 112, 114
— petrocalcic 112, 114, 128
— petrogypsic 114
— salic 113
— spodic 8, 112, 114
— sulfuric 113, 114
— surficial 8, 10, 25, 30, 64, 76, 108, 111, 112, 127, 128
Hydrolysis 43–45, 55, 56, 67, 80, 89–90, 118
Hydromorphy 11, 13, 57–58, 59–60, 66, 73, 75, 76, 77, 95, 96, 112, 113–114, 118, 120, 123, 124, 125, 127, 128
Hydrothermal
— activity 2, 19, 37, 84, 126
— adjustment 84
— alterite 2
— origin 3, 37
— percolation 4
— water 18, 19, 84

Illite 10, 13, 14, 15, 23, 28, 35, 44, 46, 47, 51, 72, 80, 82, 85, 96, 102, 103
Illuviation 10, 11, 112, 121, 129, 131
Imogolite 77
Insepic (separation) 129, 130
Intertextic (fabric) 129, 130
Iron hats, 40, 98
Isotope 20, 86, 94, 107, 110

Jarosite 32, 33, 34, 36–39, 90, 98, 103, 113

Kaolinite 10, 13, 15, 23, 35, 44, 46, 47, 48, 53, 58, 60, 61, 62, 63, 65, 67, 68, 69, 72, 73, 75, 77, 79, 82, 85, 89, 90, 103, 112, 119, 127
Karst (*also see* Paleokarst) 16, 17, 18, 65, 69–75, 88, 94, 97

Lacustrine environment 2, 19, 53, 125
— limestone (*see* Limestone)
— reef 19
Laminae 18, 28, 29, 31
Laterite 4, 39, 45, 53, 56, 67–70, 75, 77, 98, 99, 120, 122
Lattisepic (separation) 34, 130, 131
Leaching 9, 11, 13, 32, 53, 65, 69, 70, 78, 80, 108, 111, 112, 113, 120, 127, 128
Lepidochrocite 57
Leucoxene 41, 55, 83
Limonite 56, 86
Limestone
— bioclastic 97

— lacustrine 13, 14, 20, 31, 100
— palustrine 100
— siliceous 42, 53
Loess 8, 111, 120
Lutecite 40, 55, 56, 119

Mangan 75–77, 132
Manganese 75–77, 101, 113, 132
Mangrove 32, 33, 36, 38, 39, 40, 41
Marmorisation 44, 57–58, 59, 60, 66, 111, 112, 113, 123, 125, 128
Masepic (separation) 130, 131
Meteoric alteration (alterite) 2, 4, 40, 63, 67, 70, 73, 78, 84, 94, 99, 100, 126
— waters 2, 10, 18, 30, 76, 98, 102
Microcodium 23, 24, 25, 31
Micromorphology 1, 21, 24, 109–110, 128–132
Microstructure 11, 12, 31, 40, 47, 109–110, 115, 116, 117, 128–129
Mobilisation (migration) 39, 45, 67, 69, 91, 93, 98, 102, 106, 116
Molasse 8, 10–11, 14, 26, 31, 59, 75, 76, 96, 118
Monosiallitisation 89, 90, 127
Mosepic (separation) 130, 131

Neoformation 10, 13–14, 33, 34, 37, 47, 48, 61, 67, 75, 78, 81, 83, 101–103, 104, 107, 110
Nodule (nodular) 23, 28, 58, 60, 61, 63, 64, 65, 66, 67, 68, 72, 77, 109, 112

Oncolite 20, 30, 73
Opal 37, 40, 41, 44, 47, 55, 61, 68, 103
Opal-CT 40, 46, 47, 48, 55, 103
Organic matter 8–9, 11, 18, 32, 33, 53, 59, 70, 73, 75, 77, 100, 102, 103, 111, 112, 121, 122, 127, 128, 129
Oxidation 7, 8, 30, 32, 33, 39, 51, 55–77, 90, 97–99, 100, 102, 103, 104
Oysters 36, 39

Paleoalteration 1, 96, 126
Paleoclimate 16, 19, 30, 31, 56, 100, 117–119, 122–125
Paleoenvironment 1, 5, 9, 19, 39, 40, 47, 48, 49, 50, 56, 59, 67, 77, 96, 100–104, 110, 111, 122–123, 126
Paleogeography 1, 15, 16, 19, 37, 40, 50, 51, 54, 71–74, 89, 96, 122–125, 126
Paleokarst 15–18, 71–73, 94, 97
Paleomantle of alteration 79–86, 100–104, 107
Paleoprofile of alteration 3, 4, 13, 60–66, 79–86, 91–98, 104, 107, 111–115, 120, 123, 124

Paleosurface (landscape) 3, 4, 16, 25, 30, 31, 41–43, 50, 54, 55, 67, 75, 78, 84, 90, 96, 116, 117, 120, 122, 123, 124, 125, 126
Palustrine environment (milieu) 30, 53, 103
Palygorskite 10, 13, 14, 15, 30, 31, 47, 48, 103, 116, 119
Palynology 35, 36, 122
Papule 11, 12, 15, 132
Parent rock 2, 3, 10, 15, 21, 22, 31, 32, 39, 49, 55, 57, 58, 62, 63, 67, 68, 77, 85, 90–91, 106, 107, 110, 111, 117, 127, 128
Peat 112, 114, 124
Pedotubule 12, 26, 44, 59, 60, 76, 132
Pellicule 21, 22, 28, 30, 43, 49, 75
Pelosol 114
Permeability 2, 26, 104
pH 37, 47, 56, 63, 75, 76, 90, 101, 102, 103, 114, 118, 123, 124, 128
Phillipsite 103
Phytoliths 40–41
Pisolite 21, 22, 30, 58, 59, 63, 67, 68, 71, 72, 75, 77, 79, 84, 103
Planosol 128
Plasma 12, 13, 60, 129, 130, 131
Plasmic separation 34, 110, 116, 129, 130–131
Plinthite 60, 66, 67–68, 112, 114
Podzol 8, 9, 67, 84, 89, 114, 118, 120, 124, 128
Pollen 5, 36, 39
Porosity 2, 16, 17, 31, 32, 36, 42, 68, 100, 102, 104
Porphyritic fabric 129, 130
Pre-evaporite environment 15, 45–48, 50, 53–54, 103
Profile of alteration 2, 71–75, 78, 81, 83, 84–86, 90, 98, 99, 105–107, 108, 111–113, 116, 118, 121, 123, 127, 128
Pseudogley 111, 113, 128
Pyrite 32, 33, 37, 47

Quartz 8, 12, 29, 34, 40, 41–43, 45, 49, 50–51, 52, 53, 55, 56, 58, 60, 61, 63, 65, 66, 81, 85, 103, 106, 107, 119, 128
Quartzine 40, 45, 47, 48, 50, 55, 56, 119

Red beds 59, 66, 69, 98
Redox 102, 120, 124
Reduction 30, 32, 33, 39, 60, 67, 69, 75, 77, 91, 100, 104, 114, 123, 125, 127
Regolite (regolith) 2
Regression 17, 39, 69, 87, 88, 97, 99
Relief 19, 41, 69, 76, 87, 90, 98, 100, 110, 123, 125
Rendzina 25, 128

Retrometamorphism 84
Rhizolite 6
Root traces 6–7, 9, 13, 14, 21, 26, 31, 33, 34, 59, 60, 64, 109, 120, 125
Rubefaction 21, 26, 27, 57, 59, 65, 66, 76, 123, 127, 128

Sampling 8, 81, 108, 109, 110
Sandstone, Old (or New) Red 25, 26, 59, 87, 91, 96
Scales 16, 31, 54, 56, 69, 81, 90, 93, 104, 115, 116, 117, 121, 129
Sedimentary basin 62–65, 88, 92, 96, 99, 122–125
— cover/environment 59, 62–65, 69, 70, 80, 81, 84, 122, 126
— cycles 88, 120, 124, 125
— rocks 67, 81, 101, 108
— sequences, types 6–7, 15, 25, 43–45, 57, 59, 71–75, 88–89, 90–94, 96, 97, 99, 104, 105–107, 112–115, 120, 121, 123, 125
Sedimentation 6, 7, 14, 29, 48, 56, 66, 69, 71, 86, 87–99, 119–121
Sepiolite 10, 103
Sericite 10
Siderite 16, 30–31, 83, 91, 103
Siderolitic formation 48, 53, 63–67, 69–71
Sierozem 8, 114, 120, 124, 127, 128
Silan 41, 44, 46, 52, 77, 132
Silcrete 20, 37, 38, 41, 42, 43–55, 56, 84, 94, 96, 98, 99, 103, 114, 119, 120, 122
Silica (silicates) 9, 11, 23, 37, 38, 40–41, 43–55, 56, 67, 68, 69, 70, 73, 75, 77, 78, 86, 97, 104, 110, 113, 127, 128, 132
Silica precursor 49–51, 53
Siliceous accumulations 40–54, 75, 99, 119, 122
Silicification 5, 6, 30, 34, 35, 37, 39, 41, 43, 44, 45, 47, 48, 49, 50, 51–54, 55, 56, 58, 61, 63, 77, 78, 86
Silt 11
Siltan 132
Sinks (geochemical/mechanical) 71–73, 94
Skeleton soil 3, 129, 130, 131
Skelsepic separation 13, 34, 47, 110, 130, 131
Slickensides 34, 43, 47, 69, 104, 109, 128, 131
Smectite 10, 14, 15, 23, 35, 37, 47, 48, 61, 65, 77, 86, 90, 103, 119
Soil 2, 3, 69, 87–99, 113–115, 117, 119–122, 127–128, 129
— acid brown 89, 127
— acid sulfate 5, 32, 36–37, 39, 40, 103, 124
— alluvial 33, 48, 60, 64, 91–93, 127
— andic 77, 79

INDEX

— brown 64, 85, 113, 118, 124, 127
— chestnut 127
— ferrallitic 89, 90, 103, 114, 118, 120, 127, 128
— fersiallitic 77, 78, 89, 103, 113, 114, 118, 124, 127
— fossilised 1, 16, 33
— glossic (see Gloss)
— gypseous 114
— leached brown 113, 120, 127
— maroon 114, 124, 128
— podzolic ochreous 89, 114, 128
— saline 114
— tropical brown 89, 113, 128
— tropical ferruginous 113, 128
Soil age 80–81, 97, 119–122
Soil skeleton (see Skeleton soil)
Solonetz 30, 50, 114, 124, 128
Striotubule 7–8, 9, 110, 132
Stromatolite 18, 19, 20, 30
Subsidence 88, 123
Sulfate 31–39, 40, 56, 98, 101, 103, 113, 114, 128
Sulfide 9, 30, 32–39, 56, 75, 90, 91, 98, 101, 103
Sulfur 30, 34, 40, 47, 69, 75, 97, 98, 107, 116

Tectonics (tectonic activity) 18, 45, 55, 56, 87–99, 123–125
Termites 7

Time (factor) 46, 90, 95, 118, 119–122, 125
Tonstein 79
Toposequence 48, 62, 72, 90, 101, 110, 115, 123, 125
Transgression 17, 39, 60, 69, 70, 88, 95, 96, 124
Transport 2, 5–6, 20, 67, 80, 95, 100
Travertine 15, 18–20
Tubules (see Pedotubules)
Tufa 15, 18–20

Upstream zone 48, 69, 88, 90, 101, 104

Vadose 6, 16, 17, 34, 43, 59, 102, 103, 110
Vermiculite 10, 15, 82
Vertisol 10, 50, 61, 69, 89, 90, 114, 120, 124, 128
Violet zone (VLZ) 7, 30, 35, 42, 43, 44, 49–50, 79, 96, 122
Volcanic ash (rock) 6, 41, 77–79, 111, 114
Volcanism 41, 42, 77–79
Vosepic separation 13, 131

Water table 2, 13, 26, 30, 32, 43, 56, 57, 58, 59, 60, 66, 76, 79, 98, 100, 102, 105–106, 123
— phreatic 6, 10, 21, 25
Weathering 1–4, 20, 23, 42–45, 54, 61, 62, 63, 67, 68, 69, 73, 76, 77, 80, 81, 84–86, 90, 96, 99, 101, 102, 103, 104, 105–108, 110, 112, 123, 127, 128

Printed in India